The ESSENTIALS® of
REGISTERED TRADEMARK

LOGIC

D0370137

W. Kent Wilson, Ph.D.
Associate Professor of Philosophy
University of Illinois at Chicago

Research and Education Association
61 Ethel Road West
Piscataway, New Jersey 08854

THE ESSENTIALS®
OF LOGIC

Printed in the United States of America   ·

Library of Congress Catalog Card Number 98-66637

International Standard Book Number 0-87891-061-1

ESSENTIALS is a registered trademark of
Research & Education Association, Piscataway, New Jersey 08854

WHAT "THE ESSENTIALS" WILL DO FOR YOU

This book is a review and study guide. It is comprehensive and it is concise.

It helps in preparing for exams, in doing homework, and remains a handy reference source at all times.

It condenses the vast amount of detail characteristic of the subject matter and summarizes the **essentials** of the field.

It will thus save hours of study and preparation time.

The book provides quick access to the important facts, principles, statements, and programming styles in the field.

Materials needed for exams can be reviewed in summary form – eliminating the need to read and reread many pages of textbook and class notes. The summaries will even tend to bring detail to mind that had been previously read or noted.

This "ESSENTIALS" book has been prepared by an expert in the field, and has been carefully reviewed to assure accuracy and maximum usefulness.

Dr. Max Fogiel
Program Director

CONTENTS

Chapter 6
SENTENTIAL LOGIC: SYMBOLIZATION AND SYNTAX

Chapter 7
SENTENTIAL LOGIC: SEMANTICS

Chapter 8
SENTENCE LOGIC: TRUTH TREES

Chapter 9
SENTENCE LOGIC: DERIVATIONS

Chapter 10
PREDICATE LOGIC: SYMBOLIZATION AND SYNTAX

CHAPTER 1

Basic Concepts of Logic

1.1. Sentences

Logic is concerned with declarative sentences that are unambiguous and definite, and either true or false (though we may not know which).

1.1.1 Truth Values

There are two truth values: Truth and Falsity.

Every sentence to which standard logic applies has exactly one of the two truth values (that is, is either true or false, but not both). No sentence can be both true and false.

1.2 Arguments

An *argument* is a set of sentences, one of which is the *conclusion*. The remaining sentences are the *premises* of the argument, where the *premises* are taken to present evidence or reasons in support of the *conclusion*.

1.2.1 Conclusion-Indicator Words

To identify arguments in a text or conversation, it is important to understand what function each sentence in the discourse is performing. English offers some "indicator words" as clues to help identify what function sentences are serving. These serve only as a guide and do not guarantee that an argument is present. (This is because these words, like most words of English, have several different meanings.)

1

Conclusion-indicator words include: *therefore, thus, hence, so, consequently, it follows that.*

1.2.2 Premise-Indicator Words

Premise-indicator words include: *since, because, for, given that, inasmuch as, for the reason that, on account of.*

1.3 Evaluating Arguments

Arguments may be evaluated according to a variety of criteria or standards.

One such standard is the FACT standard: are all of the premises true? Is the data (evidence) offered in the premises actually, as a matter of fact, true? Are the reasons offered correct? Logic generally offers no help regarding the question of the truth or falsity of premises.

Logic is concerned with the LOGIC question: the evaluation of arguments in terms of the strength of support the premises provide for the conclusion. That is, logic develops standards for determining how strongly the premises, if all true, support the conclusion. This is an evaluation of how relevant the evidence or reasons given in the premises is to the claim made by the conclusion. Deductive logic formulates the most demanding standards by which to evaluate arguments.

1.3.1 Deductive Validity and Invalidity

An argument is (deductively) valid if and only if it is *impossible* that all its premises be true while its conclusion is false. That is, the premises of a valid argument, if they were all true, *guarantee* the truth of the conclusion; to accept all the premises and deny the conclusion would be inconsistent.

An argument is invalid if and only if it is not valid. That is, even if the premises are or were assumed (imagined) to be all true, the conclusion could still be false.

The relationship of evidential strength that premises lend to a conclusion can be seen in the following table:

Evidential Strength between Premises and Conclusion

Example of an argument that . . .	Valid (The truth of the premises guarantees the truth of the conclusion.)	Invalid (The truth of the premises does not guarantee the truth of the conclusion.)
All true premises, true conclusion	If Chicago is in Illinois, then it is in the USA. Chicago is in Illinois. Therefore, Chicago is in the USA.	Clinton was President in 1995. Therefore, Washington was the 1st US President.
All true premises, false conclusion	**By definition of "valid," none exist**	Either Bush or Clinton have been presidents. Bush was a president. Therefore, Clinton has not been President.
One or more false premises, true conclusion	Either Reagan was a President or Bush was a Vice-President. Reagan was not a President. Therefore, Bush was a Vice-President.	Bush was not a President. Therefore, Bush was a Vice-President.
One or more false premises, false conclusion	Clinton was a President and Dukakis was a President.	Bush was not a President. Therefore, Bush

| | Therefore, Dukakis | was not a Vice- |
| | was a President. | President. |

The premises of a valid argument are said to *entail* its conclusion.

The following table summarizes these relationships between truth values of premises and conclusion and the validity and invalidity of arguments.

Premises	Conclusion	Validity
all true	true	can't tell
all true	false	invalid
one or more false	true	can't tell
one or more false	false	can't tell

1.3.2 Inductive Strength and Weakness

Invalid arguments may still be good arguments when evaluated by other acceptable standards. Inductive logic develops different standards by which arguments are evaluated.

An argument is *inductively strong* if and only if it is *improbable* that all its premises be true while its conclusion is false. Inductive strength is typically measured as a real number value from zero (false) to one (certain).

An argument is inductively weak if and only if it is not inductively strong.

Examples:

The following argument is invalid, but inductively strong:
Ninety percent of restaurants in Chicago are owned by Greeks.
Lou Mitchell's is a restaurant in Chicago.
Therefore, Lou Mitchell's is owned by Greeks.

This argument is inductively strong, because given the truth of the premises, the chances are that Lou Mitchell's will be among the 90 percent of restaurants that are Greek-owned. The argument is invalid because there is some chance that Lou Mitchell's may be among the 10 percent that are not Greek-owned; so it is not *impossible* that even should the premises be true, the conclusion may still be false.

4

The following argument is valid:

100 percent (that is, all) of restaurants in Chicago are owned by Greeks.
Lou Mitchell's is a restaurant in Chicago.
Therefore, Lou Mitchell's is owned by Greeks.

Note that while the premises in fact are not true, if they were, the conclusion would have to be true. Also note that any percentage in the first premise less than 100 percent would give an invalid argument; as long as the percentage is greater than 0 percent, the argument has some inductive strength; as the percentage approaches 0 percent, the reason to accept the conclusion, given the evidence supplied by the premises, approaches no reason at all.

Symbolic logic is concerned only with deductive standards for evaluating arguments and with matters related to these standards.

1.4 Logical Properties of Sentences

1.4.1 Consistency

A sentence is *consistent* if and only if it is *possible* that it is true.

A sentence is *inconsistent* if and only if it is not consistent; that is, if and only if it is *impossible* that it is true.

Example: At least one odd number is not odd.

1.4.2 Logical Truth

A sentence is *logically true* if and only if it is *impossible* for it to be false; that is, the denial of the sentence is inconsistent.

Example: Either Mars is a planet or Mars is not a planet.

1.4.3 Logical Falsity

A sentence is *logically false* if and only if it is *impossible* for it to be true; that is, the sentence is inconsistent.

Example: Mars is a planet and Mars is not a planet.

1.4.4 Logical Indeterminacy (Contingency)

A sentence is *logically indeterminate* (*contingent*) if and only if it is neither logically true nor logically false.

5

Example: Einstein was a physicist and Pauling was a chemist.

1.4.5 Logical Equivalence of Sentences

Two sentences are *logically equivalent* if and only if it is *impossible* for one of the sentences to be true while the other sentence is false; that is, if and only if it is impossible for the two sentences to have different truth values.

Example: "Chicago is in Illinois and Pittsburgh is in Pennsylvania" is logically equivalent to "Pittsburgh is in Pennsylvania and Chicago is in Illinois."

1.5 Logical Properties of Sets of Sentences

1.5.1 Consistency (Satisfiability)

A *set of sentences* is *consistent* (*satisfiable*) if and only if it is *possible* that every sentence in the set is true; that is, if and only if it is possible that there is a situation such that each sentence of the set truly describes (some aspect of) that situation.

A *set of sentences* is *inconsistent* (*unsatisfiable*) if and only if it is not consistent; that is, if and only if for every possible situation at least one of the sentences of the set falsely describes the situation.

Where there is only one false sentence in the set of sentences, we say that the sentence is consistent (satisfiable)/ inconsistent (unsatisfiable).

Examples:

The set of sentences: {Zapata is brave; Frieda Kahlo is a gifted painter; Frieda was married to Zapata} is consistent: while the third sentence is false, it is *possible* that all three sentences are true.

The set of sentences: {Frieda was married only to Diego Rivera; If Frieda was married only to Diego, then she was not married to Trotsky; Frieda was married to Trotsky} is inconsistent. For any possible situation where the first and second sentences are true, Frieda was not married to Trotsky, and so the third sentence is false; otherwise, one of the first two sentences is false.

6

A set of sentences $\{\Box_1, \Box_2, ..., \Box_n\}$ *entails* another sentence if and only if the argument

$\Box_1$

$\Box_2$

...

$\Box_n$ /∴ Δ

is a valid argument.

A sentence, Δ, is a *logical consequence* of a set of sentences $\{\Box_1, \Box_2, ..., \Box_n\}$ $=_{\text{def}}$ $\{\Box_1, \Box_2, ..., \Box_n\}$ entails Δ.

The informal definitions given above have been expressed in terms of what is *possible* and *impossible*. This indicates that the basic concepts that were defined are systematically interrelated. However, the concepts of *possibility* and *impossibility* need to be made clearer.

A set of sentences is *technologically impossible* (at a given time, to a given culture) if and only if the situation they describe is not attainable given the technology available to that culture at that time.

For example, it is presently technologically impossible for a train to exceed 750 mph in prolonged travel.

A set of sentences is *physically impossible* if and only if the situation they describe is contrary to the laws of nature (physics, chemistry, biology, etc.).

For example, it is physically impossible for a train to go faster than the speed of light.

A set of sentences is *logically impossible* if and only if the situation they describe is contrary to the laws of logic.

For example, it is logically impossible for a train to go 20 mph while standing perfectly still.

The relationship between these kinds of possibility is shown in the following graphics:

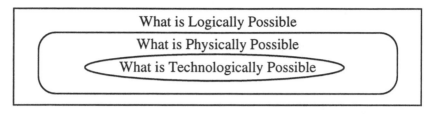

Figure 1-1

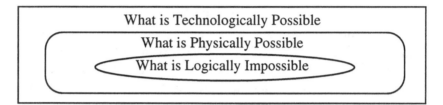

Figure 1-2

One task for logic is to make the concepts of logical possibility and impossibility more precise and to introduce methods to determine what is possible and what is not possible.

CHAPTER 2

Meaning and Definition

2.1 Extension and Intension of Terms

The *extension* (or *denotation*) of a term is the object, or set of objects, that it applies to (i.e., that it is true of). For example, the extension of the expression "The author of *Tom Sawyer*" is Mark Twain; the extension of "city of North America" is the set of cities consisting of Chicago, Montreal, Mexico City, etc.

The *intension* (or *connotation*) of a term is the properties that are jointly necessary and sufficient for the correct application of the term. For example, the intension of "triangle" is the properties of being a plane closed figure and having three straight sides.

Two expressions may have the same extension but have different intensions. For example, the expressions "the author of *Tom Sawyer*" and "the novelist who created *Huckleberry Finn*" apply to exactly the same object, Mark Twain, but the intensions of the two expressions differ: a person could understand one of the expressions but not understand the other. Another example: the terms "unicorn" and "centaur" apply to the same set of objects, namely to no objects at all, yet the words have different intensions, for they are not synonyms. The extension of these two words is said to be empty.

There are no expressions that have the same intension but different extensions, since intension determines extension: the intension of an expression is the criterion used to decide what objects are in its extension.

2.2 Types of Definitions According to Purpose

Definitions are groups of words that explain or specify the meaning of a given expression, either as the person giving the definition intends to use it or as that expression is commonly used in a language.

The expression to be defined is the *definiendum*. The expression that does the defining is the *definiens*.

For example: In the definition

triangle = $_{def}$ plane closed figure bounded by three straight sides

"triangle" is the definiendum and "plane closed figure bounded by three straight sides " is the definiens.

In general, the purpose of giving a definition is to facilitate communication by explaining the meanings of expressions. Definitions may serve several more specific functions in explaining the meaning of an expression. Some of these follow.

2.2.1 Stipulative Definition

A stipulative definition is a definition whose purpose is to explain what the person giving the definition intends to mean by the expression. This usually involves coining a new word to stand for something but it may involve assigning a new meaning to a word in common use. The purpose of a stipulative definition is usually to abbreviate a longer expression or to introduce a convenient or brief way to talk about something (for example, a new product or phenomenon).

Examples:

N^n =$_{def}$ N times itself n times; thus, 10^3 is 10 x 10 x 10;

Tigon =$_{def}$ the offspring of a male tiger and a female lion.

A stipulative definition does not attempt to capture the meaning of a term in common use; it is an arbitrary assignment of meaning to an expression for convenience. The assignment may be more or less appropriate or useful, but no question of correctness or incorrectness arises for stipulative definitions.

2.2.2 Lexical Definition

A lexical definition is a definition whose purpose is to explain the conventional or commonly understood meaning of an expression. Lexical definitions aim to report what people mean by the expression defined. Thus, a lexical definition can be more or less accurate, correct or incorrect. Dictionary definitions are lexical definitions.

Ambiguous words have more than one meaning. A lexical definition lists the different meanings an expression has and thereby serves to expose such ambiguities. Thus, the adjective

Fair: 1. Of pleasing appearance 2. Light in color, especially blond 3. Free of clouds or storms 4. Free of blemishes or stains 5. Having or exhibiting a disposition that is free of favoritism or bias; impartial 6. Consistent with rules, logic, or ethics

These are just a few of the definitions listed in the *American Heritage Dictionary*.

2.2.3 Precising Definition

An expression is *vague* when there is no exact condition specifying whether the expression applies or does not apply. A characteristic of vague expressions is that they give rise to borderline cases where the application of an expression to a particular object is neither clearly true nor clearly false. For example, the lexical definition of "intoxication" is "to be stupefied or excited, due to the action of a chemical substance such as alcohol."

The purpose of a precising definition is to set more precise (though more or less arbitrary) limits on the range of application of an expression that is vague in its ordinary meaning, the point of which being that such precision is required for some special purpose (such as for legal reasons). As such, a precising definition involves a stipulative element within the bounds of the ordinary meaning of the expression which would be given by its lexical definition.

The lexical meaning of "intoxication" is too vague to serve general legal purposes. In the Illinois statutes, an intoxicated person is defined as one "whose mental or physical functioning is substantially impaired due to the presence of alcohol in the body." This also is

11

vague, and for the purposes of the motor vehicle code in identifying drunk drivers, "intoxication" is defined as having a blood alcohol concentration of 0.10 or higher. These are both precising definitions.

2.2.4 Theoretical Definition

The purpose of a theoretical definition is to explain the meaning of an expression in terms of a particular theory (scientific, philosophical, or otherwise) in which the expression plays a significant role. For example, "heat" may be defined as "mean kinetic energy of molecules," thereby related to kinetic molecular theory.

2.2.5 Persuasive Definition

The purpose of a persuasive definition is to give the meaning of an expression in such a way as to prejudice the attitudes of others positively or negatively toward the objects denoted by the expression. The purpose of doing this is so that a claim will appear more plausible or less plausible than it would if the expression were used with its conventional meaning. For example, someone might define

"liberal": a person who favors ever more powerful governments completely controlling most aspects of our daily lives and increasingly taxing our incomes.

This is rhetorically devious, because if one accepts the definition, one concedes more than is appropriate given the common meaning of the word "liberal."

2.3 Types of Definitions According to Method of Defining

Various methods may be used to define expressions. The following is not an exhaustive list of all such methods, but a survey of some of the more important means of constructing definitions. These methods may be classified into extensional and intensional definitions.

2.4 Extensional (Denotative) Definitions

An extensional definition explains the meaning of an expression by indicating the extension of the definiendum (that is, by indicating

12

the objects to which the expression applies). Obviously, this method is limited by the fact that in some cases there are simply too many objects or kinds of objects in the extension of the definiendum to list (e.g., "rational number"); in other cases there are no objects in the extension of the definiendum (e.g., "unicorn"); and in some cases, many of the objects in the extension of the definiendum have no names and so cannot be easily listed. Since two terms can have the same extensions but different intensions (e.g., "creature with a heart" and "renate"), an extensional definition may not enable a person to apply the definiendum to new instances.

2.4.1 Enumerative Definition

An enumerative definition of a term gives the extension of the term. Strictly speaking, it must give the complete extension of the definiendum, but often only a partial extension is given. The extension may be given singly, where each object denoted by the definiendum is listed:

Canadian Province: Alberta, British Columbia, Manitoba, New Brunswick, Newfoundland, Nova Scotia, Ontario, Price Edward Island, Quebec, Saskatchewan

or collectively (sometimes called *definition by subclass*), where all the kinds of things denoted by the definiendum are listed:

Noble gas: Argon, Helium, Krypton, Neon, Radon, Xenon

An example of an enumerative definition which only gives a partial listing of the extension of the definiendum is:

Angiosperm: oaks, maples, elms, walnuts, hickories, and the like.

(Problem: what else is included? In fact, not pines, spruces or firs)

2.4.2 Ostensive Definition

An ostensive definition gives the meaning of a term by pointing to or displaying the extension of the definiendum. For example, an ostensive definition of "red" might consist of saying "Red is *that* color" while pointing to a red color swatch, or "The Sears Tower" means *that* (pointing to the building).

2.4.3 Recursive (Inductive) Definition

This is a kind of definition that is encountered in logic, mathematics, and other formal studies. A recursive definition proceeds in three stages: (1) the base clause characterizes a subclass of the extension of the definiendum; (2) the induction or recursive clause gives a rule for determining the remaining objects in the extension of the definiendum by relating any such object to an object the definiendum already applies to; (3) the closure clause states that the definiendum applies to no other objects. For example, "ancestor" can be defined recursively as follows:

(1) A person's parents are the person's ancestors;

(2) A parent of a person's ancestor is a person's ancestor;

(3) Nothing else is a person's ancestor.

Here, clause (1) makes one's parents one's ancestors. Clause (2), applied to one's parents, places one's grandparents among one's ancestors. Clause (2), applied to one's grandparents, makes one's great-grandparents one's ancestors. And so on.

2.5 Intensional (Connotative) Definitions

An intensional definition explains the meaning of an expression by specifying its intension; that is, by giving the properties of objects that are necessary and sufficient for the expression to apply correctly to those objects. Intensional definitions may take several different forms.

2.5.1 Synonymous Definition

A synonymous definition is one in which the definiens is a single word having the same meaning as the definiendum (hence the name "synonymous definition"). Examples:

"Physician" means doctor

"Retiform" means net-like

"Saccharase" means sucrase

2.5.2 Operational Definition

An operational definition specifies an operation or series of operations or experimental procedures to be performed to determine whether or not the definiendum applies in a particular case. For example:

14

to determine whether the term "acid" applies to a particular liquid, insert a strip of blue litmus paper into the liquid; the liquid is an acid if and only if the litmus paper turns red.

A testing operation has been specified: inserting a strip of blue litmus paper into the liquid; a specific test result determines whether the definiendum applies to the liquid: it does if the litmus paper turns red, it does not if the paper does not turn red.

2.5.3 Contextual Definition

A contextual definition explains the meaning of the definiendum by explaining the meaning of any larger expression in which the definiendum may occur. For example:

"A unless B" means if it is not the case that B, then A.

The logical symbol for "or," $\lor$, can be contextually defined in terms of the logical symbols for "and " (&) and "not" (~):

$A \lor B =_{def} \sim (A \& B)$.

This definition allows any occurrence of "$\lor$" in a context to be replaced by its definiens.

2.5.4 Definition by Genus and Species

Logicians use "genus" and "species" as relative terms. By "genus" logicians mean the broader class of items relative to some class of objects (the "species"), and by "species" logicians mean a smaller subclass of objects contained within the genus.

Definition by genus and species involves two steps: (1) the extension of the definiendum is located within some broad class of objects, the genus (not all objects in this class are members of the extension of the definiendum, however); (2) some feature of the members of the extension of the definiendum is indicated (the *difference* or *specific difference*) that distinguishes them from the remaining objects in the genus. For example:

Square: rectangle that has equal sides (genus: rectangle; specific difference: having equal sides)

Human: rational animal (genus: animal; specific difference: rationality)

2.6 Rules For Definitions

The following rules are guidelines to framing and evaluating definitions. Not every rule applies to every kind of definition.

First, **A Definition Should Make Sense and Conform to the Standards of Proper Grammar**. This includes spelling, punctuation, and sentence structure.

2.6.1 A Definition Should State the Essential Attributes of the Members of the Extension of the Definiendum

The idea is that a proper definition should capture what is *necessary* for an object to be in the extension of the definiendum, rather than merely what is *accidental* or *coincidental* about the objects in its extension. For example, the definition of "human" as featherless biped is criticized on the grounds that being featherless and bipedal are not essential properties of human beings. Four ways in which a definition can be seen to fail in capturing the essence is if it violates one of the following two rules.

A Definition Must Be Neither Too Broad Nor Too Narrow. This rule is aimed primarily toward lexical definitions. A definition is *too broad* if objects that have the properties specified by the definiens are not in the extension of the definiendum. On the customary or intended use the definiendum does not apply to some objects that the definiens allows it to apply to. For example, the definition

"Essay" means a literary composition

is too broad, since it would place poems, novels, and plays in the extension of "essay," when its customary or intended meaning excludes them.

A definition is *too narrow* if objects in the extension of the definiendum fail to have the properties specified by the definiens and as a result would not be included in the extension of the definiendum if the definition were accepted. On the customary meaning of the definiendum, it applies to objects that the definiens does not allow it to apply to. For example, the definition

"Sandal" means a shoe with a leather sole fastened to the foot with thongs or straps is too narrow, since the sole of a sandal need not be made of leather.

16

A definition can suffer from being both too broad and too narrow. For example,

"Knowledge" means true belief about an event

our definition is too narrow, in that knowledge applies to states as well as events, and it is too broad since it allows lucky guesses (true but unwarranted beliefs) as knowledge.

A Definition Should Not Be Negative When an Affirmative Definition is Possible. This rule requires that a proper definition explain what the definiendum *does* stand for rather than what it *does not* stand for. For example, the definition

"Scalene" means a triangle that is not equilateral or isosceles

is needlessly negative and thereby fails to capture the essential properties of scalene triangles, namely that of having all three sides unequal.

A Definition Should Be Literal. A figure of speech such as a metaphor does not give the essential properties of the extension of the definiendum. To define "dog" as man's best friend may tell us something useful about the relation between dogs and people, but it hardly tells us what properties to look for to determine whether something is a dog. Similar remarks hold for other figures of speech such as irony and hyperbole.

A Definition Should Be Evaluatively Appropriate. A definition should not introduce evaluative language unless the definiendum is itself an evaluative expression. Thus sarcasm and facetious language are not appropriate for definitions intended for serious work. This rule rejects persuasive definitions and "loaded" definitions which attempt to introduce an evaluative claim gratuitously without presenting evidence or argument for the claim. This rule is violated when instead of reporting on the meaning of the definiendum, the definiens editorializes about the extension of the definiendum in order to influence opinions or attitudes about the extension.

2.6.2 A Definition Should Not Be Circular

A circular definition is one in which the definiens uses the very word to be defined. Proposing a circular definition obviously defeats the purpose of giving a definition: to explain the meaning of the definiendum.

Example: "Cause" means something that causes an effect. This characterization of circularity concerns singly or intrinsically circular definitions. Circularity can also occur in more complicated ways.

Definition often is useful in systematically presenting a subject matter (geometry being a classic example). In that circumstance, several definitions when taken together may form a circular chain of definitions. To avoid this, we may adopt the special version of the rule against circularity that follows.

Definitions Must Be Mutually Non-Circular. As definitions are added to a systematic presentation of a subject, each new definition must not create a circle. It is easy to appreciate what is wrong with this in the case of a simple two-definition circle:

"Cause" means something that produces an effect.

"Effect" means something produced by a cause.

Note that neither definition by itself is circular, but plugging the meaning of "effect" into the definiens of "cause" we obtain the completely uninformative

"Cause" means something that produces something that is produced by a cause.

2.6.3 A Definition Must Be Mutually Consistent With Other Definitions That Have Been Previously Given

As definitions are introduced in a systematic study, care must be taken to ensure that an inconsistency does not arise, either because of an inconsistency with previous definitions or because of an inherent inconsistency within the definition itself. For example, if, in a system of arithmetic, we define division contextually by

"$x/y = z$" means $x = y \cdot z$

we can easily derive a contradiction by letting $y = 0$ and x and z be any two numbers. (The contradiction will be that $x = 0$, where $x > 0$.)

2.6.4 A Definition Must Be Clear (Unambiguous, Not Excessively Vague, Not Obscure)

A proper definition should not have an ambiguous definiens; otherwise the definition will give more than one meaning to a given

18

entry for the definiendum. Where a definiendum is itself ambiguous, then distinct entries should be listed for each distinct meaning.

While some vagueness is unavoidable in many cases, a definiens that is excessively vague will fail to explain what objects the definiendum applies to. For example, the definition

"Communism" means a bad form of government

does not give any clear understanding what kinds of government the word "communism" applies to.

CHAPTER 3

Categorical Propositions

3.1 Constituents of Categorical Propositions

Categorical propositions in *standard form* can be thought of as sentences that have one of the following four forms, letting "S" and "P" stand for *subject term* and *predicate term*, respectively:

Form of Categorical Proposition	Letter Name	Example
All S are P	A	All robins are birds.
No S are P	E	No robins are birds.
Some S are P	I	Some robins are birds.
Some S are not P	O	Some robins are not birds.

Subject terms and predicate terms denote classes of objects. The words "all," "some," and "no" are *quantifiers*. The quantifiers indicate how much of the subject class is included (or not included) in the predicate class. The words "are" and "are not" are called the *copula*: they link the subject and predicate terms in the proposition. It is convenient to refer to these forms by the letter names indicated.

3.2 Quality, Quantity, and Distribution

3.2.1 Quality

A categorical proposition has *affirmative quality* if it affirms class membership, as in the A and I forms. Thus

A: All S are P   states that all members of S <u>are</u> members of P; that is, the class S is included in the class P (positive or affirmative).

I: Some S are P   states that some member of S <u>is</u> a member of P; that is, the class S and the class P have at least one member in common (positive or affirmative).

A categorical proposition has *negative quality* if it denies class membership, as in the E and O forms. Thus

E: No S are P   states that <u>no</u> member of S is a member of P; that is, the class S is <u>excluded</u> from the class P, the two classes are <u>disjoint</u> (negative).

O: Some S are not P   states that there is at least one member of S that is <u>not</u> a member of P; that is, the class S has at least one member that is <u>not</u> a member of P (negative).

3.2.2 Quantity

The quantity of a categorical proposition is either *universal* if it makes a claim about all members of the subject class or *particular* if it makes a claim about some (at least one) member of the subject class. The A and E propositions have universal quantity and the I and O propositions have particular quantity. (An E proposition is universal because it states that <u>all</u> members of the S class are excluded from the P class.)

Particular propositions are understood in a narrow and strict way as meaning just that there exists at least one member of the S class that is/is not a member of the P class. An I proposition is <u>not</u> understood as stating that some S are P and some are not. An O proposition is <u>not</u> understood as stating that some S are not P but some S are P.

3.2.3 Distribution

A term is *distributed* in a proposition if the proposition states something about every member of the class denoted by the term; otherwise it is undistributed.

21

The concept of distribution is not intuitive, and the characterization above is vague. A better explanation is an extensional one:

The subject term of A and E propositions is distributed; the predicate terms of E and O propositions is distributed. No other terms are distributed in categorical propositions.

SUMMARY

Propositional Forms	Letter	Quality	Quantity	Terms Distributed
All S are P	A	Affirmative	Universal	S
No S are P	E	Negative	Universal	S, P
Some S are P	I	Affirmative	Particular	none
Some S are not P	O	Negative	Particular	P

3.3 Traditional Square of Opposition

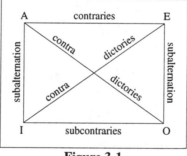

Figure 3-1

The traditional square of opposition represents logical relations that were believed to hold among the various categorical propositions.

Contradictory propositions cannot both be true and cannot both be false; that is, of a pair of contradictory propositions, exactly one is true and exactly one is false. The traditional square of opposition represents the A and O as contradictories and the E and I as contradictories.

Contrary propositions are propositions that cannot both be true but can both be false. The traditional square represents the A and E as contraries.

22

The relation of *subalternation* holds that the truth of the A or E proposition entails the truth of the respective I or O proposition. That is, the arguments

A /∴I and E /∴O

are valid. Equivalently, the falsity of the I and O propositions entails the falsity of the corresponding A and E propositions.

The correctness of these logical relations of the traditional square of opposition depends crucially on the assumption of *existential import* of subject terms in the A and E propositions. This assumption states that the extension of the subject term is not empty, that there are Ss. Where this assumption fails, all the logical relations represented by the traditional square of opposition fail to hold except for the relation of contradictories between A and O propositions and E and I propositions.

Suppose that we want to reason about unicorns. Since there are no unicorns (the class of unicorns is empty), the I and O propositions

I: Some unicorns are mammals

O: Some unicorns are not mammals

are both false, since each says that *there are unicorns*, the I proposition saying that some are mammals and the O saying that some are not mammals. Since both are false, they cannot be subcontraries, as the traditional square claims.

Since both I and O are false, the contradictory of each must be true. So the

A: All unicorns are mammals

and the

E: No unicorns are mammals

are both true. So although the traditional square represents them as contraries, they cannot be contraries.

Finally, since the A is true and the I is false in this example, the subalternation relation cannot obtain between them as is true and similarly for the E and O propositions.

The traditional interpretaion of A propositions is: All S are P and there are As; the interpretaion of E propositions is: No S are P and there are Ss.

23

3.4 Modern Square of Opposition

The modern square does not make the assumption of existential import. As a result, the only logical relation among categorical propositions that is retained in the modern square is the relation of contradictoriness, represented by the propositions on the diagonals of the square:

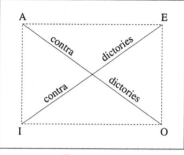

Figure 3-2

The modern interpretation of A propositions is: There does not exist an S that is not a P. The modern interpretation of E propositions is: There does not exist an S that is a P. The modern interpretation of the I and O propositions are the same as the traditional interpretation.

3.5 Venn Diagrams

The extension of a term can be represented by a circle. Since each categorical proposition has two terms, two circles are required to represent each categorical proposition. The universe is represented by a box enclosing any and all circles, though it is common to omit this.

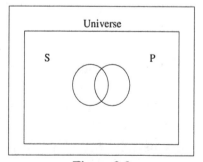

Figure 3-3

24

Logical relationships between the two terms, as expressed in a categorical proposition, are represented by shading to show that a region is empty and by placing an "X" in a region to show that it has at least one object in it. The four forms of categorical proposition are represented as follows.

A: All S is P

I: Some S is P

E: No S are P

O: Some S are not P

3.6 Conversion, Contraposition, Obversion

Conversion, contraposition, obversion, and are three rules of "immediate inference." They are called "immediate inferences" because only one premise is involved. The typical use of each of these rules transforms the sentence it is applied to into a logically equivalent sentence. Hence either sentence can be validly inferred from the other.

3.6.1 Conversion

This rule permits one to switch the S and P terms in E propositions and in I propositions. Using "::" to represent logical equivalence, we have

No S are P :: No P are S   No snakes are dogs :: No dogs are snakes

Some S are P :: Some P are S   Some dogs are pets :: Some pets are dogs

Conversion is <u>not valid</u> for A and O propositions. Using ":/:" to express that the two propositions on either side are <u>not</u> logically equivalent:

25

All S are P :/: All P are S

All dogs are pets :/: All pets are dogs.

Some S are not P :/: Some P are not S

Some animals are not dogs :/: Some dogs are not animals.

3.6.2 Contraposition

The complement of a term T is a term that applies to all and only those things that T does not apply to. The complement is standardly formed in logic by prefixing "non-" to the original term. Thus, the complement of some term T is "non-T." The complement of the complement of T is non-non-T, which is logically equivalent to T.

Contraposition is similar to conversion. The subject and predicate terms are switched (as in conversion), but in switching them, each is replaced by its complement.

Contraposition only applies correctly to A and O propositions.

A: All S are P :: All non-P are non-S

All dogs are animals :: All non-animals are non-dogs (i.e., anything that is not an animal is not a dog)

O: Some S are not P :: Some non-P are not non-S.

Some animals are not snakes :: Some non-snakes are not non-animals (i.e., some things that aren't snakes are not things that aren't animals).

Contraposition does not apply to E and I propositions:

No S are P :/: No non-P are non-S.

No snakes are dogs :/: No non-dogs are non-snakes. (A python is a non-dog, but it is a snake.)

Some S are P :/: Some non-P are non-S.

Some politicians are honest :/: Some non-honest thing is a non-politician (A convicted alderman is non-honest but is a non-politician.)

Note that the mechanics of the rule are simple. Switch subject and predicate terms, prefix each by "non-," and leave everything else

26

the same. But the rule applies only to A and O propositions; it doesn't apply to E and I propositions.

3.6.3 Obversion

The mechanism used in applying the rule is this:

(1) Change the quality.

(2) Replace the predicate term by its complement.

This rule applies to all categorical propositions.

A: All S are P :: No S are non-P	All vipers are snakes :: No vipers are non-snakes
E: No S are P :: All S are non-P	No snakes are dogs :: All snakes are non-dogs
I: Some S are P :: Some S are not non-P	Some snakes are vipers :: Some snakes are not non-vipers
O: Some S are not P :: Some S are non-P	Some snakes are not dogs :: Some snakes are non-dogs

The correctness of any of these rules can be verified using Venn Diagrams. The Venn Diagram for a given categorical proposition is the same as the Venn Diagram for the proposition that results from applying the rule.

3.7 Translation Hints

Many sentences of English are not in the form required for categorical propositions. The logic of categorical propositions is not as restricted as this would make it appear, because many sentences which are not of the proper form can be paraphrased into sentences that are of the proper form. The relations represented by the square of opposition and the rules of immediate inference can help here.

For example:

Not all politicians are crooks.	Some politicians are not crooks. (Contradictory)

27

| Some non-gases are not incompressible | Some compressible substances are not gases (Contraposition) |

There is no mechanical method that will generally apply for constructing a paraphrase of any sentence, though there are some additional guidelines that apply in particular cases.

3.7.1 Temporal Reference

The verb in a categorical proposition must be the copula. It does not matter whether it is singular or plural, so "is" / "is not" do not have to be paraphrased. But sentences containing tensed verbs must be paraphrased. Some examples will give you the idea.

Some Greeks were physicians	Some Greeks are people who were physicians.
All dogs would rather bark than bite.	All dogs are creatures that would rather bark than bite.
Everything will perish.	All things are things which will perish.
Some US President will be a female.	Some female person will be a US President.

Some female person is a person who will be a US President

Note that in the last example, we apply conversion and then paraphrase the result. If we tried the simpler paraphrase "Some US President will be a female," we obtain a sentence that seems to say that a male President will undergo a change in sex, which is unlikely to have been the intended meaning.

3.7.2 Verbs Other Than The Copula

In a great many English sentences, of course, other verb forms are used besides "to be." These also admit of straightforward paraphrasing, as the following examples show:

| All Caribou migrate seasonally. | All caribou are animals that migrate seasonally. |
| Some plants lose their leaves in winter. | Some plants are vegetables that lose their leaves in winter. |

28

3.7.3 Terms Without Nouns

Subject and predicate terms must be noun phrases in standard categorical propositions. So when we have sentences with predicate adjectives or other improper terms, we have to paraphrase them finding an appropriate noun phrase.

Some daises are yellow.	Some daises are yellow flowers.
All lions are carnivorous.	All lions are carnivorous animals.
No spitting is allowed.	No act of spitting is an act which is allowed.

3.7.4 Singular Propositions

A singular proposition is one which contains a proper name instead of a subject term. For examples:

Socrates is a Greek.

There is no completely satisfactory way to paraphrase such sentences into a categorical form. What is generally recommended is to paraphrase them on the model of

All things identical with Socrates are things that are Greek.

A negative singular sentence is to be paraphrased along the lines of

Socrates is not a Greek.	No thing identical with Socrates is a thing which is Greek.

One problem is that the two paraphrases are only contraries, not contradictories. But "Socrates is a Greek" and "Socrates is not a Greek" are contradictories: exactly one is true and the other is false.

The treatment with Venn Diagrams can be simpler. In the case of particular propositions, we represent them by placing an "x" in a region of intersecting circles because we don't know which entity exists that is in the region, only that there is at least one. But for singular propositions, such as

Harry is a private.

we know which entity it is, so we can use a proper name representing the entity in our Venn Diagram. As names we use lower case letters from "a" to "v" to represent individual objects. The sentence above can be represented as follows:

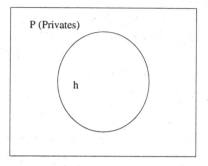

Figure 3-4

3.7.5 Adverbs

Temporal adverbs, such as "when," "where," "anytime," "always," "never," etc. quantify over times, and can be paraphrased as follows:

Anytime I see him he's smiling.	All times when I see him are times when he is smiling.
She always cooks ham on New Year's day.	All times that are New Year's day are times when she cooks ham.
Sheldon never eats beans.	No time is a time when Sheldon eats beans.

Spatial adverbs, "where," "wherever," "anywhere," "everywhere," "nowhere," etc. quantify over places:

Anywhere Jose goes, people are glad to see him.	All places that Jose goes are places where people are glad to see him.
Nowhere in Chicago will you find an honest alderman.	No places in Chicago are places where you will find an honest alderman.

As usual, care is necessary. A sentence like

Men are sometimes wicked.

often is incorrectly paraphrased as

Some men are wicked.

rather than as

Sometimes are times when men are wicked.

30

3.7.6 Nonstandard and Unexpressed Quantifiers

In English there exists expressions of quantity that are not among the three standard quantifiers recognized in categorical logic. These expressions are called *nonstandard quantifiers*. In some cases, we can do pretty well representing these English quantifiers in terms of our standard quantifiers; in other cases it's a stretch; and in some cases it's so great a stretch that the result is inadequate to the logic of the English sentence. It is important to try to grasp what could plausibly be said.

Some English quantifiers that can't be adequately translated for syllogistic arguments are "many," "most," "few," "exactly one," "at least two," etc. You can get a glimpse of this with "few," which conveys more than "some":

Few men are good generals.

Some men are good generals; some men are not good generals.

The sentence cannot be adequately paraphrased by a single categorical proposition or even by this pair of categorical propositions.

To illustrate the importance of understanding what the English sentence says and why there is not a simple rule for paraphrasing every sentence, consider:

<u>A</u> chicken is a domesticated animal.	<u>All</u> chickens are domesticated animals.
<u>A</u> badger is not a domesticated animal.	<u>No</u> badger is a domesticated animal.
<u>A</u> domesticated animal is living in my house.	<u>Some</u> domesticated animal dwells in my house.

Some quantity expressions that can often be satisfactorily paraphrased are the following:

<u>A few</u> officers are gentlemen.	Some officers are gentlemen.
<u>Certain</u> women are great explorers.	Some women are great explorers.

Cognates of "all": "every," "each," "any," cognates of "some": "at least one," "a," "an," and cognates of "no" "none," "nothing," can be easily paraphrased.

Pronouns such as "who," "whoever," "anyone," etc. quantify over people (or perhaps some pets). Pronouns such as "whatever," "what," "anything," "everything," etc. quantify over "things" if no more specific kind of obvious.

Many English sentences with no overt quantity expression can be paraphrased as a categorical proposition. These are called *unexpressed quantifiers*. Here again, it is crucial to reflect on the meaning of the sentence to be paraphrased. Some examples:

Penguins swim.	All penguins swim.
Babies are helpless.	All babies are helpless persons.
Babies are in the bedroom.	Some babies are persons in the bedroom.

Exclusive propositions are those that involve the English quantifiers "only," "none but," "none except," and "no ... except." These quantifiers <u>exclude</u> one class from another. For example:

Only men are allowed into the shower area of the Bull's dressing room.

This is true. Now ask: are <u>all</u> men allowed into the shower area of the Bull's dressing room? Most men are not allowed into <u>any</u> part of the Bull's dressing room. So the sentence is not saying that all men are allowed, it is saying that <u>only</u> men are allowed; i.e., all those allowed into the shower area are men (anyone who is allowed must be male). So

Only S are P

is translated as

All P are S.

Similarly for "none but":

None but S are P   All P are S

None but the brave deserve the fair   All persons who deserve the fair are brave persons.

"The only" generally means "all." It is the opposite of exclusive propositions. Thus,

The only S are P   All S are P

The only reliable cars are Fords.   All reliable cars are Fords.

Exceptive propositions are not to be confused with exclusive propositions. These are propositions expressed by sentences containing the words "All except S are P," or "All but S are P."

These are translated as

No S are P and All non-S are P.

Thus

All except people nominated for office are eligible to vote. All people who haven't been nominated are eligible to vote; and no nominated person is eligible to vote.

Note that exceptive propositions cannot be paraphrased satisfactorily as categorical propositions; they must be paraphrased as two categorical propositions.

CHAPTER 4

Categorical Syllogisms

4.1 Standard Form, Mood, and Figure of Syllogisms

4.1.1 Categorical Syllogisms

A *categorical syllogism* is an argument consisting of two premises and a conclusion such that the conclusion is not entailed by either premise alone (by an immediate inference, for example). Each premise must be a categorical proposition (or paraphrasable into a categorical proposition—for the remaining discussion, we assume that all paraphrasing has been completed). For a categorical syllogism to have a shot at being valid, there must be exactly three terms occurring in the three statements making up its premises and conclusion; so each term will occur twice, once in each of two propositions. When more than three terms occur, the argument commits the categorical Fallacy of Four Terms. Each term that occurs in the syllogism must be used in the same sense throughout the argument. (This requirement prevents a fallacy of equivocation from occurring, which is, in effect, a version of the Fallacy of Four Terms.)

An example:

Only men are rational.   (= All rational beings are Human.)
No female is a man.   (= No female being is a male being.)
Therefore, no female is rational.

There is an obvious equivocation on the term "man"; the argument commits the Fallacy of Four Terms.

34

The subject term of the conclusion is called the *minor term*; the predicate term of the conclusion is called the *major term*. The premise in which the major term occurs is called the *major premise* and the premise in which the minor term occurs is called the *minor premise*. The third term, common to each premise, is called the *middle term*.

4.1.2 Standard Form of Categorical Syllogisms

A categorical syllogism is in *standard form* when its major premise is listed first, then its minor premise, and finally its conclusion:

Standard Form of Categorical Syllogisms

Major Premise

Minor Premise

∴ Conclusion

4.1.3 Mood of Syllogisms

Since each sentence occurring in a categorical syllogism is a categorical proposition, each is one of the four forms A, E, I, O. The mood of a syllogism is indicated by the three letters representing the forms of the major premise, minor premise, and the conclusion in that order. For example,

No P is M	E
All S is M	A
∴ Some S is not P	O

has the mood EAO.

4.1.4 Figure of Categorical Syllogisms

The *figure* of a syllogism represents the pattern of occurrences of the middle term in the syllogism. There are four figures which are illustrated leaving out quantifiers and copulas (letting "S" represent the minor term and "P" represent the major term):

	1st Figure		2nd Figure		3rd Figure		4th Figure	
Major premise	M	P	P	M	M	P	P	M
Minor premise	S	M	S	M	M	S	M	S

35

The four figures can be remembered in terms of the "W"-shaped figure, keeping in mind that the S and P terms in the 2nd and 3rd figures are inside the angles:

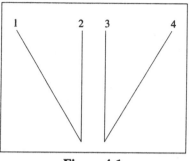

Figure 4-1

The combination of mood and figure gives a complete classification of the logical forms of categorical syllogisms. The standard way of representing a categorical syllogism is by writing its mood followed by its figure; for example, AII-3 is a syllogism of the form

All M is	P
Some M is	S

∴ Some S is   P

Each categorical syllogism consists of three sentences, and each sentence can be one of four moods. That gives 4 x 4 x 4 = 64 possible moods. There are four figures, so there are 64 x 4 = 256 different categorical syllogisms. It is possible to list the valid ones: there are only 15 valid categorical syllogisms. If the traditional interpretation is adopted, there are four to nine more, depending on exactly what assumption of existential import is adopted. It is no longer necessary to memorize the valid forms; the valid syllogisms can be determined either by the rules of the syllogism or by Venn Diagrams.

4.2 Rules of the Syllogism

The following rules characterize valid categorical syllogisms in the following sense: if a rule is violated, the syllogism is invalid; and if no rule is violated, the syllogism is valid.

<u>Rule 1</u> In a valid syllogism, the middle term must be distributed in at least one premise. A syllogism that fails to satisfy this rule commits the <u>Fallacy of Undistributed Middle</u>. Example of the fallacy[1]:

> All chimpanzees are primates.
> All humans are primates.
> _____
> ∴ All humans are chimpanzees.

<u>Rule 2</u> In a valid syllogism, a term that is distributed in the conclusion must be distributed in a premise. A syllogism that fails to satisfy this rule commits the <u>Fallacy of Illicit Major</u> if the major term is distributed in the conclusion but not in the major premise, or the <u>Fallacy of Illicit Minor</u> if the minor term is distributed in the conclusion but not in the minor premise. Examples of the fallacies:

<u>Illicit Major</u>	<u>Illicit Minor</u>
All dogs are animals.	No lions are tigers.
Some cats are not dogs.	All lions are cats.
∴ Some cats are not animals.	∴ No cats are tigers.

<u>Rule 3</u> A valid syllogism cannot have two negative premises. A syllogism that violates this rule commits the <u>Fallacy of Exclusive Premises</u>. An example of the fallacy:

> No snakes are Chicago aldermen.
> No snakes are politicians.
> _____
> ∴ No politicians are Chicago aldermen.

<u>Rule 4</u> A valid syllogism that has an affirmative conclusion cannot have a negative premise (a negative premise requires a negative conclusion). A syllogism that violates this rule commits the <u>Fallacy of Drawing an Affirmative Conclusion from a Negative Premise</u>. An example of the fallacy:

> No chimps are fish.
> All sharks are fish.
> _____
> ∴ Some sharks are chimps.

<u>Rule 5</u> A valid syllogism cannot have a particular conclusion when both premises are universal. A syllogism that violates this rule commits the <u>Fallacy of Existential Import</u>. On the traditional inter-

pretation of universally quantified sentences, valid syllogisms may violate this rule. Example of the fallacy:

All mammals are vertebrates.
All centaurs are mammals.

∴ Some centaurs are vertebrates.

Two additional rules that can be derived from the basic rules above which are very easy to apply.

Rule 6 No valid syllogism can have two particular premises.

Rule 7 A valid syllogism that has a negative conclusion must have a negative premise.

Rule 4 and Rule 7 together imply that a valid syllogism cannot consist of only one negative sentence; a valid syllogism will have either two negative sentences or none.

A syllogism may commit more than one fallacy. For example, any argument of the form

No M is P
All M is S

∴ All S is P

commits the Fallacy of Illicit Minor and the Fallacy of Drawing an Affirmative Conclusion from a Negative Premise.

1 Examples of fallacies will be arguments having both premises true but the conclusion false, and thus be invalid arguments, and will violate the rule illustrated.

Chapter 5

Venn Diagrams

5.1 Venn Diagrams

Venn Diagrams are graphical representations of categorical propositions and their logical interrelations.

5.1.1 Constructing Venn Diagrams

Using Venn Diagrams to test syllogisms for validity adapts the method introduced to represent categorical propositions as categorical syllogisms. Since a syllogism contains three terms, a Venn Diagram for that syllogism will consist of three overlapping circles:

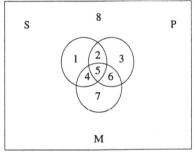

Figure 5-1

Each area represents potential information. If we place an "x" in an area, that represents that there is at least one object in that area. Thus, for example, placing an "x" in region 7 represents an object that is M but neither S nor P. Shading a region represents the fact that there is nothing in that region: it is empty of objects. Thus shading regions 4 and 5 represents that there is no object that is both

S and M. Obviously, a region can't both be empty and have at least one object in it; that would be contradictory.

When dealing with diagrams involving three circles, a problem arises with diagramming particular (and singular) sentences. Let's consider the case of an I proposition: "Some S are P" in the context of a diagram for a syllogism:

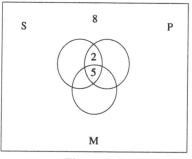

Figure 5-2

The problem we face is where to put the "X" in the S circle: in region 2? in region 5? in both? The answer is that the statement "Some S are P" does not tell us into which region to put the "X"; it does not tell us whether the S is a P but not an M (region 2), or whether it is an S, a P, and also an M (region 5). We can't include information in our diagram that we are not given in the premises. We represent this situation by putting an "X" in region 2 and another "X" in region 5, but we connect them with a bar, indicating that an object is in one or the other or both regions, the premise does not tell us which. When the other premise is diagrammed, this indeterminacy may be resolved. But whether it is resolved or not, the method will not be affected. Note that "Xs" connected by a bar gives <u>less</u> information than simply an "X" in a region; it is more informative to know that something is in region 2 than it is to know that there is something either in region 1 or in region 2.

<u>Strategy hint:</u> To eliminate indeterminacy as much as possible, diagram universal statements before diagramming any particular (or singular) statements. If we knew that one of the two regions above was empty, then we would know that the "X" would have to be placed in that part of the S-P intersection that was not empty.

Example:

All M are P

Some S are M

―――――――――

∴ Some S are P

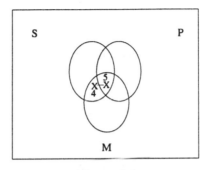

Figure 5-3

If we begin by diagramming the minor premise, we must place "Xs" connected with a bar in regions 4 and 5, as shown in Figure 5-3. When the major premise is diagrammed, region 4 and region 7 are shaded, and so empty. The only place for there to be an object, as required for the minor premise to be true, is in region 5. See Figure 5-4. Notice that if the major (universal) premise had been diagrammed first, only one non-empty region, region 5, would have remained in which to place an "X" when diagramming the minor premise.

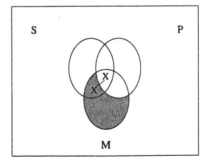

Figure 5-4

5.1.2 Using Venn Diagrams to Evaluate Categorical Syllogisms

To use Venn Diagrams to evaluate syllogistic arguments involves two steps:

Step 1 Diagram the premises of the syllogism.

Step 2 Inspect the Venn Diagram to determine whether it represents the information expressed by the conclusion. That is, ask: "Does anything need to be added to the diagram in order that the conclusion be represented as true?" If "No," the argument is valid; if "Yes," the argument is invalid.

Example of a valid syllogism: see the example and Figure 5-4 above. Note that to diagram the conclusion of the syllogism would require an "X" in region 2 or in region 5 (or both); but an "X" in region 5 represents the conclusion as being true.

Example of an invalid syllogism:

All M are P

No S are M
_____

∴ No S are P

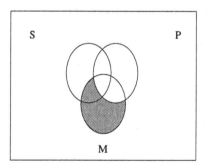

Figure 5-5

In Figure 5-5, the major premise has been diagrammed (regions 4 and 7 shaded [i.e., empty]). In Figure 5-5, we diagram the minor premise (regions 4 and 5 shaded [i.e., empty]). The conclusion requires that regions 2 and 5 be shaded (empty), but that information is not represented in Figure 5-6. Hence the syllogism is invalid.

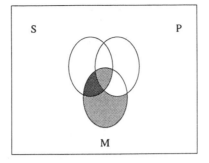

S   P

M

Figure 5-6

5.2 Enthymemes

An *enthymeme* is an incompletely stated syllogism; either a premise has been omitted or the conclusion has not been stated, usually because it is believed to be too obvious to need explicit expression. When the mission statement is supplied, the result is a categorical syllogism. (The term "enthymeme" is also used for any incomplete expressed argument, not just incomplete syllogisms.) To evaluate, supply the missing statement, and proceed as usual.

Example: Those animals are canines, since they are dogs. Missing premise: All dogs are canines.

5.3 Sorites

A *sorites* is a series of categorical propositions that can be organized into a sequence of categorical syllogisms, where the conclusion of each syllogism in the sequence (except the last) is used as a premise in the next syllogism of the sequence. For example:

1. All S are F

2. All F are M

3. All M are P

∴ All S are P

(1) and (2) yield the intermediate conclusion "All S are M." This together with (3) yields the conclusion. There can be any number of categorical propositions that form a sorites.

The task in evaluating sorites can be divided into two sub-tasks:

43

<u>Step 1</u> Put the sorites into standard form. Different texts adopt different standard forms. (It is somewhat arbitrary which form of several equally good ones is adopted as standard.)

<u>Step 2</u> Evaluate the individual arguments of the sequence and the overall argument. This will be done by constructing a series of Venn Diagrams.

5.3.1 Standard Form For Sorites

A sorites is in standard form iff:

(1) All component propositions are in categorical form, each term occurring exactly twice (i.e., each term occurs in each of exactly two premises).

(2) The predicate term of its conclusion occurs in the first premise.

(3) Each successive premise has a term in common with the preceding premise.

The object of putting a sorites in standard form is to make it easier to evaluate the argument.

5.3.2 Evaluating Sorites Arguments

(1) In order to evaluate the sorites argument, we must take pairs of categorical propositions as premises and derive a conclusion. The part of propositions we take must contain exactly three terms, so that categorical logic applies. Given our standard form, the first two propositions will contain three terms, so we can construct a Venn Diagram for them.

(2) Infer a conclusion from the Venn Diagram. There are four guidelines for drawing intermediate conclusions.

    (i)   The intermediate conclusion will be a categorical proposition that is distinct from those used as premises.

    (ii)   The intermediate conclusion must contain an occurrence of a term that occurs in the next premise to be used.

    (iii)   The predicate term of the conclusion of the syllogism must be preserved; i.e., any intermediate conclusion must contain an occurrence of that term.

44

(iv) The middle term of the syllogism must not occur in the conclusion you draw, so this intermediate conclusion must relate the term that occurs only once in the premises to the predicate term of the conclusion of the sorites argument.

(3) Taking that intermediate conclusion as a premise, together with the next premise on the list, a Venn Diagram is constructed. Note that if you have proceeded correctly, this should be a categorical syllogism. Step (2) is repeated, and a new conclusion is generated.

(4) This process is repeated until the conclusion of the sorites argument is the conclusion of the argument being tested.

(5) If a chain of valid arguments can be constructed in this way, then the sorites argument is valid. If it <u>cannot</u> be constructed, it is not. <u>Note:</u> not just "I can't construct a chain of arguments," but that no such chain can be constructed.

Consider the following sorites argument (from Lewis Carroll's *Symbolic Logic*):

All babies are illogical persons.	All B are I
Nobody is despised who can manage crocodiles.	No D are M
Illogical persons are despised persons.	All I are D
Therefore, no baby can manage crocodiles.	No B are M

In standard form, this sorites becomes

(1) No D are M

(2) All I are D

(3) All B are I

∴ No B are M

Stage 1. Take (1) and (2) and draw conclusion:

We draw the intermediate conclusion "No M are I." See Figure 5-7.

45

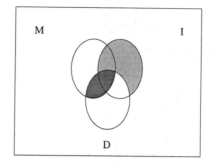

Figure 5-7

Stage 2. Take (3) and the intermediate conclusion "No M are I" and diagram the syllogism

No M are I

All B are I

∴ No B are M

In diagramming the premises of this second argument, the conclusion has been represented. See Figure 5-8. Hence the sorites is valid.

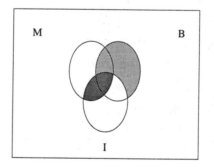

Figure 5-8

CHAPTER 6

Sentential Logic: Symbolization and Syntax

6.1 Sentential Logic

Sentential logic (also called "propositional logic" and "statement logic") concerns logical relations among sentences that depend on the presence of one or more occurrences of five sentence operators: negation ("not"), conjunction ("and"), disjunction ("or"), conditional ("if ... then"), and biconditional ("if, and only if," commonly abbreviated "iff").

Simple or atomic sentences are represented by *sentence letters.* These represent sentences which for the purposes of logical analysis are taken to have no relevant logical structure. Texts vary in whether to choose as sentence letters all or some upper case or lower case letters of the alphabet. In what follows in this book, sentence letters will be taken to be any upper case letters from the alphabet. (In the chapters on predicate logic, this will be qualified slightly.)

A simple language for representing logical structures can be constructed using the resources of sentence letters and operators. This formal language is often called "SL" (for "symbolic language for sentence logic").

To characterize a language, we must at least:

- specify a vocabulary—the set of symbols to be used in the language;

- specify a grammar or syntax—give the rules for putting vocabulary items together to obtain acceptable complex units such as sentences;

- specify a semantics for the language—give the rules indicating what various simple or complex items of the language mean.

In the remainder of this chapter, the vocabulary and grammar of a sample SL will be presented, and some guidelines for representing sentences of English in SL will be given.

6.2 The Vocabulary of SL

The vocabulary of SL consists of sentence letters, sentential operators (or connectives), and punctuation.

<u>Atomic Sentences of SL</u>

The *atomic sentences* or sentence letters of SL are upper case letters from "A" to "Z," with or without numerical subscripts.

<u>(Sentential) Connectives</u>:

Symbol of SL	Concept expressed
~	"not"; negation
&	"and"; conjunction
V	"or"; disjunction
⊃	"if ... then ___"; conditional
≡	"... if and only if ___"; biconditional

<u>Punctuation:</u>

(,) ; [,] ; {, }; etc.

The five sentence operators or connectives listed above are represented by different symbols in different texts. Some of the variants are listed below, using "A" and "B" as sample sentences, to illustrate the grammar.

Operator	Quasi-English example	Symbols
Negation	Not-A	~A; ¬A; -A; $\overline{A}$
Conjunction	A and B	A & B; A ∧ B; A • B; AB

48

Disjunction	A or B	A VB; A V B
Conditional	If A then B	A ⊃ B; A → B; A => B
Biconditional	A if and only if B	A ≡ B; A <=> B

6.3 Grammar of SL

In specifying the grammar of SL it helps to have some special symbols. We let □, Δ, O, and ◊ stand for any sentences of SL. <u>Note: These geometric shapes are NOT themselves sentences of SL; they "stand for" any sentences of SL.</u>

The grammatical rules of SL are sometimes called "Formation Rules" because they are rules for forming more complex sentences (molecular sentences) out of atomic sentences.

<u>Formation Rules of SL</u>

1. Atomic sentences are sentences of SL.

2. If □ is a sentence, then ~□ is a sentence of SL.

3. If □, Δ are sentences of SL, then:

    a. (□ & Δ) is a sentence of SL;

    b. (□ V Δ) is a sentence of SL;

    c. (□ ⊃ Δ) is a sentence of SL;

    d. (□ ≡ Δ) is a sentence of SL.

4. Nothing else is a sentence of SL.

Note: This is a recursive definition of "sentence of SL."

Negation is a *unary connective*: take any sentence, put a "~" in front of it to form the negation of the original sentence. The other four connectives are *binary connectives*: take any two sentences, place the binary connective between them, and put the resulting molecular sentence in parentheses to obtain a new sentence.

When "&" is placed between two sentences, the resulting molecular sentence is a *conjunction*; each of the two constituent sentences is called a *conjunct*. When "V" is placed between two sentences, the resulting molecular sentence is a *disjunction* (in some texts called an *alternation*); each of the two constituent sentences is called a *disjunct*. When "⊃" is placed between two sentences, the molecular sentence that results is a *conditional*; the constituent sentence to the left of the "⊃" is called the *antecedent* of the conditional and the constituent sentence to the right of the "⊃" is called the *consequent* of the conditional. When "≡" is placed between two sentences, the resulting molecular sentence is called a *biconditional*.

6.4 Additional Grammatical Concepts

Every molecular sentence has exactly one *main connective* or *main logical operator*. Since atomic sentences don't have any connectives occurring in them, they don't have a main connective. The main connective of a molecular sentence is the connective introduced by the last formation rule applied in the construction of the sentence from its constituents.

The *atomic constituents* of a sentence of SL are all the atomic sentences that occur in it (that is, all that are constituents of it).

A *literal* is an atomic sentence or a negation of an atomic sentence; thus, "B" is a literal, as is "~ F."

The *immediate sentential constituents of a molecular sentence* are:

Molecular sentence	Main sentential connective	Immediate sentential constituents
negation	~	sentence negated
conjunction	&	each conjunct
disjunction	V	each disjunct
conditional	⊃	antecedent, consequent
biconditional	≡	left and right constituents

The *subformulas (subwffs; sentential constituents) of an SL sentence* ☐ are:

a.   ☐ itself;

b.   all the immediate subwffs of ☐, the immediate subwffs of its immediate subwffs, &c.

These additional grammatical concepts can be more formally defined by means of a recursive definition, as follows:

50

1. If $\square$ is an atomic sentence, no connectives occur in $\square$; so $\square$ does not have a main connective, and $\square$ has no immediate sentential constituents other than itself.

2. If $\square$ has the form $\sim\bigcirc$, then the main connective of $\square$ is "$\sim$," and is the only immediate sentential constituent of $\square$.

3. If $\square$ is of the form

    (a) $\bigcirc$ & $\triangledown$, then the main sentential connective of $\square$ is "&," and $\bigcirc$ and $\triangledown$ are the immediate sentential connectives of $\square$;

    (b) $\bigcirc$ V $\triangledown$, then the main sentential connective of $\square$ is "V," and $\bigcirc$ and $\triangledown$ are the immediate sentential connectives of $\square$;

    (c) $\bigcirc \supset \triangledown$, then the main sentential connective of $\square$ is "$\supset$," and $\bigcirc$ and $\triangledown$ are the immediate sentential connectives of $\square$;

    (d) $\bigcirc \equiv \triangledown$, then the main sentential connective of $\square$ is "$\equiv$," and $\bigcirc$ and $\triangledown$ are the immediate sentential connectives of $\square$.

6.5 Basic Translation Hints

Translation from English to SL often requires careful paraphrasing of the English sentence before finally rendering it into a sentence of SL. There are some regularities, but even these must be treated carefully, for there are exceptions.

Negations English cognates: "not," "no," "non-," "un-," "it is not the case that."

In English, the grammar of negation is more complicated than it is in SL:

Socrates is not foolish.   :: $\sim$F   (F: Socrates is foolish.)

Socrates didn't write
  the *Republic*.   :: $\sim$W   (W: Socrates wrote the *Republic*.)

Not everyone is happy.   :: $\sim$E   (E: Everyone is happy.)

When in doubt, paraphrase the English sentence using "it is not the case that" before translating into SL.

51

<u>Conjunctions</u> English cognates: "and," "but," "although," "however," "both," "also," "moreover."

"And" may occur between parts of speech besides sentences, as in

Susan and Angela are athletes.	::	S & A (S: Susan is an athlete; A: Angela is an athlete.)
Yoko admires and respects Lincoln.	::	A & R (A: Yoko admires Lincoln; R: Yoko respects Lincoln.)

Care must be taken, however. The following sentences can't be translated into SL as conjunctions:

Bob and Carlo are brothers.	::	? B & C says "Bob is a brother and Carlo is a brother"; but are they each other's brother? "B&C" doesn't say that, only that each is someone's brother.
Donald and Ivana hate each other.	::	? D&I says "Donald hates each other and Ivana hates each other. (Doesn't even make sense.)

In these sorts of cases, the best we can do is translate the sentences by sentence letters.

<u>Disjunction</u> English cognates: "or," "either ... or ...," "unless," "at least one" (in uses where the alternatives have been listed).

"Or" may occur between constituents that are not sentences. Again, careful thought is required to identify cases where the translation into SL can be made by a disjunction.

The simplest rule for translating "unless" into SL is to translate it as "or":

A unless B:: A V B

Raf won't forgive you unless you apologize:: Raf won't forgive you or you will apologize.

There are alternatives that sound more plausible, but they are more complicated. For example:

A unless B :: A if not B   (Then translate this using the condi-

52

tional and negation — see below.)

I'll come to the party unless Al is coming :: If Al isn't coming, then I'll come to the party.

Similarly for "Unless A, B" :: "If not A, then B."

Conditionals The paradigm form for a conditional in English is a sentence of the form

If □, then Δ.

English cognates: "if □ then Δ," "Δ if □," " Δ provided that □," "Δ assuming that □," " Δ on the condition that □," "□ only if Δ."

Only if is troublesome. "□ only if Δ" can be translated into SL as "□ ⊃ Δ." More plausibly, it may be translated by "~Δ ⊃ ~□," which is logically equivalent to "□ ⊃ Δ":

My pulse exceeds 150 only if I exercise. :: If I don't exercise, my pulse doesn't exceed 150.

In general, "if" in English is followed by the antecedent of a conditional "except" when it is preceded by "only"; in that case, it signals the consequent of a conditional.

Unless "□ unless Δ" :: "□ if not Δ" :: "If not Δ, then □" :: ~Δ ⊃ □

"~Δ ⊃ □" is logically equivalent to "□ VΔ", which is why "unless" can be translated as "or."

"If A then B" does not mean "If A then B and if not A then not B."

Biconditionals English cognates: "if, and only if," "when, and only when," "just in case."

Some Common Sentences That Can Be Translated Using Just "V", "&", and "~"

Neither ... nor ...:: Not (either ... or ...)

Neither A nor B :: ~(AVB) which is logically equivalent to ~A & ~B.

The Cubs will win neither the Series nor the pennant. :: ~(PVS) which is logically equivalent to ~P & ~S.

Not both:

Not both A and B :: ~ (A & B) which is logically equivalent to ~AV ~B.

53

Ann: Roy, Tommy failed his algebra and the spelling quiz today.

Roy: Surely, not both. :: ~ (A & S)

which is logically equivalent to ~A∨ ~ S :: Either he didn't fail algebra or he didn't fail spelling.

6.6 Ambiguity and Punctuation

Punctuation is used to prevent ambiguity from arising. Sentences can be ambigous not only by containing an ambiguous word but also because of grammatical structure. For example, the English sentence "Old men and women will be evacuated" is ambiguous: is it old men and old women or old men and all women who will be evacuated? In algebra, an expression written "$a + b \times c$" is ambiguous between: (a) adding a with b and multiplying the sum by c; (b) adding a to the product of b times c. In SL, the expression "A & B ∨C" is ambiguous: is its main connective a conjunction or a disjunction?

The grammar specified above prevents such ambiguities from occurring in SL. The simplest way of writing the grammatical rules (given above) can give rise to some unnecessary parentheses. These are introduced because, when the expression is completed, the parentheses are not required, but they would be required if one were to add more to that expression. The principal place where excess parentheses are generated by our grammatical rules is:

- Outermost parentheses around the complete SL sentence. These can always be erased (or you can omit to write them). Thus, there is no harm in writing "(P & Q)" as "P& Q."

The grammar of negation makes the negation sign operate on the smallest sentence following it consistent with the punctuation. Thus, in the sentence "~A & B," all that is negated is the sentence "A." To negate the conjunction, we have to write "~(A & B)." The following three sentences each make different statements:

~A & B ~(A & B)   ~A & ~B

The first says "A isn't true but B is"; the second says "it's not the case that both A and B are true"; and the third says "A isn't true and B isn't true." In the second sentence, the parentheses are essential; in the other two sentences, inessential outer parentheses have been omitted.

CHAPTER 7

Sentential Logic: Semantics

7.1 Truth Values and Truth Functions

The logic to be presented is a two-valued logic. The semantics recognizes two *truth values: truth* and *falsity.*

The connectives of SL are *truth-functional*: this means that the truth value of any complex sentence formed using any of the connectives of the vocabulary depends entirely on the truth values of the simpler sentences which make up the complex sentence. Since any sentence of SL is constructed ultimately from atomic sentences and connectives, the truth value of any complex (molecular) sentence of SL will depend on nothing other than the truth values of the atomic sentences that occur in it.

7.2 Truth Values Assignments

The semantics of SL is given by considering any assignment, $\mathcal{A}$, of truth values to sentence letters and stating what must be the case for any given sentence of SL, $\Box$, to be true under an assignment $\mathcal{A}$. This way of giving a semantics reflects the fact that the very minimum a semantics for a language should provide is the conditions under which sentences are true or false.

Let $\mathcal{A}$ be any assignment of truth values to sentence letters.

1. Atomic sentences are either true (t) or false (f) under $\mathcal{A}$.
2. If $\bigcirc = \sim\Box$, then **TV** $(\bigcirc)$ = t under $\mathcal{A}$ iff **TV** $(\Box)$ = f under $\mathcal{A}$.

55

3. If $\bigcirc = (\square \ \& \ \triangle)$, then **TV** $(\bigcirc) = $ t under $\mathcal{A}$ iff **TV** $(\square) = $ **TV** $(\triangle) = $ t under $\mathcal{A}$.

4. If $\bigcirc = (\square \lor \triangle)$, then **TV** $(\bigcirc) = $ t under $\mathcal{A}$ iff **TV** $(\square) = $ t under $\mathcal{A}$ or **TV** $(\triangle) = $ t under $\mathcal{A}$, or both.

5. If $\bigcirc = (\square \supset \triangle)$, then **TV** $(\bigcirc) = $ t under $\mathcal{A}$ iff either **TV** $(\square) \neq $ t under $\mathcal{A}$ or **TV** $(\triangle) = $ t under $\mathcal{A}$, or both.

6. If $\bigcirc = (\square \equiv \triangle)$, then **TV** $(\bigcirc) = $ t under $\mathcal{A}$ iff **TV** $(\square) = $ **TV** $(\equiv)$ under $\mathcal{A}$.

This is a recursive definition of "true on a truth value assignment."

Each clause of the definition represents a "semantic valuation rule." Valuation rules can be expressed in the form of a statement or graphically, in the form of a characteristic truth table for each of the sentential connectives.

Clause (1) expresses the fact that assignments of truth values (**TV**s) to sentence letters can be made in any way that suits us, subject to certain very minimal conditions:

• Every sentence letter (occurring in the problem we're working on) is assigned a **TV** by an assignment.

• No sentence letter can be assigned more than one **TV** by an assignment.

7.2.1 Negation: ~

Clause (2) above represents the valuation rule for negations:

<u>Semantic Valuation Rule</u> : A *negation* is true if and only if (iff) the sentence negated is false; otherwise the negation is false; that is, the negation of a sentence has the "opposite" **TV** of the un-negated sentence. So if $\square$ is true, then ~$\square$ is false; but if $\square$ is false, then ~$\square$ is true.

Characteristic Truth Table for Negation

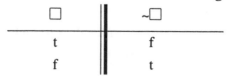

$\square$	~$\square$
t	f
f	t

Note: Letting ☐ be any sentence of SL, it follows from the semantics for negation that ☐ and ~☐ cannot both be true and cannot both be false; at least one is false and at least one is true. A sentence and its negation are contradictories.

7.2.2 Conjunction: &

Clause (3) represents the valuation rule for conjunctions:

Semantic Valuation Rule: A conjunction is t iff both its conjuncts are t; otherwise it is f.

Characteristic Truth Table for Conjunction

☐	Δ	☐ & Δ
t	t	t
f	t	f
t	f	f
f	f	f

Note: Let ☐ be any sentence of SL. Then according to our semantical valuation rules for "&" and for "~," no sentence of the form "☐ & ~☐" can be true; any such sentence is logically false in SL. Note also that no sentence of the form "~(☐ & ~☐)" can be false; any such sentence is logically true in SL.

7.2.3 Disjunction: V

Clause (3) represents the valuation rule for disjunctions:

Semantic Valuation Rule: A disjunction is t iff at least one disjunct is t; otherwise it is f; that is, a disjunction is true iff at least one of its disjuncts is true.

Characteristic Truth Table for Disjunction

☐	Δ	☐ V Δ
t	t	t
f	t	t
t	f	t
f	f	f

Note 1: Let □ be any sentence of SL. Then any sentence of the form "□ V~□" must be true according to the semantical valuation rules for "V" and for "~." Any sentence of the form

"~(□ V~□)" then, must be logically false.

Note 2: The semantics for SL makes a sentence of the form "□ V Δ" true when at least one of its disjuncts is true. It follows that the disjunction is true when both disjuncts are true. So "□ V Δ" means "□ or Δ (or both)." This represents the *inclusive* sense of "or."

Many logicians believe that there is an exclusive sense of "or" in English. The best current theory is that there is no such sense. However, "exclusive or" can be represented in SL easily. Let's use the symbol "▼" as a connective expressing "exclusive or." Its grammar will be like that of any of the binary connectives.

Semantic Valuation Rule for Exclusive "or": An exclusive disjunction is t iff exactly one of its disjuncts is t. I.e., the **TV** of an exclusive disjunction, (□ ▼ Δ) = t iff either **TV** (□) = t and **TV** (Δ) = f or **TV** (□) = f and **TV** (Δ) = t; otherwise, the exclusive disjunction is f.

Characteristic Truth Table For Exclusive "Or"

□	Δ	□ ▼ Δ
t	t	f
f	f	t
f	t	t
f	f	f

An "exclusive or" can be defined using the vocabulary of SL as follows:

$$□ ▼ Δ =_{def} (□ V Δ) \& \sim (□ \& Δ)$$

"A ex-or B" means "Either A or (inclusive) B, but not both."

When using truth tables or when calculating the truth value (**TV**) of a compound sentence given some **TV**s for its immediate constituents, it is not always necessary to enter the **TV**s of all the atomic sentences or immediate constituents directly beneath every occurrence of them.

It is not always necessary to write their **TV**s under all the constituents to compute the **TV** of a compound. In particular:

58

A conjunction with one or more false conjuncts is false, no matter what the **TV** of the other conjunct is.	A disjunction with one or more true disjuncts is true, no matter what the **TV** of the other disjunct is.

7.2.4 Conditional: ⊃

Clause (5) represents the valuation rule for conditionals:

<u>Semantic Valuation Rule</u>: A conditional, □ ⊃ Δ, is true iff either its antecedent, □, is false or its consequent, Δ, is true (or both). More simply, a conditional is false iff its antecedent is true and its consequent is false; otherwise it is true.

Characteristic Truth Table for Conditionals

□	Δ	□ ⊃ Δ
t	t	t
f	t	t
t	f	f
f	f	t

The table displays the fact that the only circumstance when a conditional of SL is false is when its antecedent is true and its consequent is false.

Applying the semantical rule for conditionals can be simplified in certain cases. The rule tells us, in effect, that a conditional with a true antecedent is true no matter what the truth value of its consequent. It also tells us that a conditional with a true consequent is true no matter what the truth value of its antecedent.

A conditional with a false antecedent is true.	A conditional with a true consequent is true.

7.2.5 Biconditional: ≡

Clause (6) gives the valuation rule for biconditionals:

<u>Semantic Valuation Rule</u>: A biconditional, □ ≡ Δ, is true iff □ and Δ have the same truth value; i.e., either □ and Δ are both true or □ and Δ are both false.

Characteristic Truth Table for Biconditionals

□	Δ	□ ≡ Δ
t	t	t
f	t	f
t	f	f
f	f	t

7.2.6 Truth Functions And Non-Truth-Functional Connectives

SL is a truth functional language. This is shown by the fact that in the characteristic truth tables for each connective, each cell has exactly one **TV** entered. Given the **TVs** of all the atomic sentences occurring in a given sentence, the **TV** of the molecular sentence can be calculated according to the Valuation Rules for the language.

Natural languages such as English contain non-truth-functional connectives (NTFCs); for example, English conditionals are rarely used truth-functionally. The main point here is to alert you to the fact that there are NTFCs in English, to indicate what some NTFCS are, and to explain how to deal with them with the available resources. Generally, the best we can do is to represent sentences that contain NTFCs as atomic sentences.

Common non-truth-functional connectives include:

Subjunctive conditionals (in fact, most English conditionals, even indicative conditionals) are NTFCs. The best we can do is to represent them as conditionals or to represent them as sentence letters (i.e., as atomic).

Modals: *may, maybe, might, can, could, should, must, possibly, necessarily, probably, impossible, improbable, ought* are all NTFCs.

Verbs of Propositional Attitude: *believe, know, want, wish, seems, imagines, hopes, seeks, sees, realizes,* etc. are all NTFCs.

7.3 Truth Values Assignments and Possibility

In Chapter 1, the basic concepts of logic were defined in terms of possibility and impossibility. These definitions suffer from certain shortcomings:

- Vagueness in the concept of "possibility";

- No obvious way to apply the definitions in particular cases.

These shortcomings are corrected here.

Possibility is explained in terms of *Truth Values assignments* (*TV assignments*): a possibility is simply an assignment of truth values to atomic sentences. A distinct **TV** assignment (a distinct possibility) is represented by each row in a truth-table. The totality of the rows of a properly constructed truth-table represents the totality of all (relevant) **TV** assignments (possibilities). A sentence is possible if it is represented as being true on one or more rows of a truth-table. A sentence is impossible if it is represented as true on no row of a truth-table.

A truth value (**TV**) assignment assigns truth values to all the atomic sentences of SL. But for any given problem, there will be a finite number — usually a fairly small number — of atomic sentences that occur in the sentences that comprise the problem; all other atomic sentences of SL will be irrelevant to the problem. As a result, in dealing with a given problem we need to be concerned only with those parts of the **TV** assignments that are assignments of **TV**s to those atomic sentences that actually occur as constituents of the sentences that make up the problem.

The rules of valuation determine the **TV** of molecular sentences of SL given that **TV**s have been assigned to their constituents (and ultimately, to their atomic constituents). As a result of applying the valuation rules after **TV**s have been assigned to atomic constituents, the **TV** of any molecular sentence occurring in a problem can be computed. In a truth-table, each row represents a **TV** assignment to the (relevant) atomic sentences occurring in the problem. For each row, we can then compute the **TV** of the sentences (molecular or atomic) occurring in the problem by using the valuation rules to compute their **TVs** given the **TVs** of their constituents. We will let script letter A represent any such **TV** assignment.

7.3.1 Truth and Falsity of a Sentence on a Truth Value Assignment

For any given **TV** assignment $\mathcal{A}$ of **TVs** to atomic sentences of SL, a sentence $\square$ is <u>true on $\mathcal{A}$</u> iff $\square$ takes the value t on that assignment of **TVs** to atomic sentences occurring in $\square$ (as determined by the valuation rules).

For any given **TV** assignment $\mathcal{A}$ of **TVs** to atomic sentences of SL, a sentence $\square$ is <u>false on $\mathcal{A}$</u> iff $\square$ takes the value f on that assignment of **TVs** to the atomic sentences occurring in $\square$ (as determined by the valuation rules).

7.4 Truth Tables

7.4.1 The Construction of Truthtables

The overall structure of a truth table

Left-hand side	Right-hand side

The sentence or sentences that make up the problem being solved are listed across the top of the right-hand side of the truth table. There are two stages in the construction of a truth-table:

    (i) Construction of the left-hand side;

    (ii) Construction of the right-hand side.

Construction of the left-hand side of a truth table begins by listing across the top of the left-hand side all the atomic sentences that occur in all the sentences of the problem.

There are two problems in the construction of the left-hand side of a truth-table: (1) determining how many possibilities must be represented; and (2) finding a systematic way of listing all the possibilities so none are overlooked. (This can happen by repeating one set of assignments to atomic sentences so two assignments are really the same.)

    (1) <u>Determining how many possibilities there are for a given number of atomic sentences</u>

62

Each atomic sentence can be true or false, independently of what **TV** any other atomic sentence has been given. Thus:

Method to calculate the total combinations of Truth Values:

For N distinct atomic statements, there are 2^N distinct combinations of truth values (2 **TV**s, N statements); that is, for N distinct sentence letters, there are 2^N possibilities. Consequently, the left-hand side of the truth table will have to contain 2^N rows.

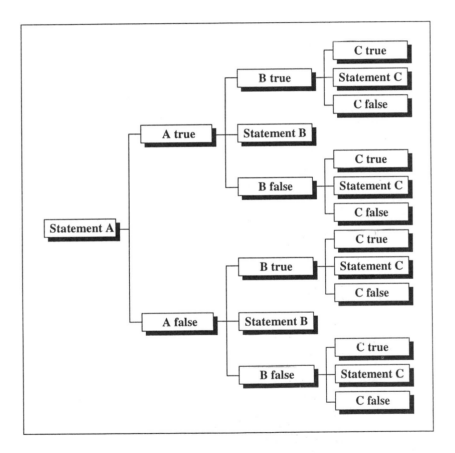

Figure 7-1

Figure 7-1 shows that with one sentence letter there are two possibilities; with two sentence letters there are four possibilities;

and with three sentence letters there are eight possibilities. The pattern displayed in Figure 7-1 illustrates how the number of possibilities doubles with the addition of each sentence letter. The following table lists the number of possibilities for truth tables having up to six sentence letters on the left-hand side.

Number of atomic sentences	Number of possibilities
1	2
2	4
3	8
4	16
5	32
6	64

(2) <u>A systematic way of listing all the distinct possibilities</u>

Each row must represent a <u>distinct</u> possibility. That means that each row must differ from every other row in at least one entry. Where many rows are required to list all the possibilities, it can be difficult to be sure that each row you write is different from all the previous rows that have been entered. A systematic way of listing the possibilities is desirable. There are several methods, the main difference being whether one begins to write in truth values in the left-most column of the left-hand side of the truth table, or begins at the right-most column of the left-hand side of the truth table. Suppose the left-most column is selected to take the first entries. Then "t"s and "f"s are alternated as many times as there are possibilities calculated. If the truth table requires eight possibilities, for example, then alternate "t" and "f" so that the left column is eight rows high:

Step 1

A	B	C	
t			
f			
t			
f			
t			
f			
t			
f			

Next, move one column to the right, and write twice as many "t"s as you wrote in the previous column before writing an "f"; in this case, write two "t"s and then two "f"s. Follow this pair of "t"s with a pair of "f"s. Repeat until the column is filled. (Refer to Stage 2 below.)

Stage 2

A	B	C	
t	t		
f	t		
t	f		
f	f		
t	t		
f	t		
t	f		
f	f		

Stage 3

A	B	C	
t	t	t	
f	t	t	
t	f	t	
f	f	t	
t	t	f	
f	t	f	
t	f	f	
f	f	f	

Move one column to the right, and write twice as many "t"s as appeared before an "f" in the previous column; in this case, four "t"s. Follow these by an equal number of "f"s. Repeat until the column is filled. (Refer to Stage 3 above.) Continue this process until the columns under all the atomic sentences are filled in.

The sentences that occur in the problem to be solved are written across the top when constructing the right-hand side of a truth table.

To complete the right-hand side of the truth table, fill in the columns under the sentences listed across the top by applying the Valuation Rules in a step by step way until all the rows of all the columns are filled with **TVs**. In each row, the **TVs** assigned on the left-hand side to the atomic sentences are used in determining the **TVs** of the molecular sentences for that row.

7.5 Truth Values Assignments, Truth Tables, and Semantic Properties

The semantic properties previously defined in terms of possibility and impossibility can now be defined in terms of truth value assignments and characterized by truth tables. Each row of a truth table is a *Truth Value assignment*.

The basic concept for the semantics of SL is that of *satisfaction of a sentence by a TV assignment*.

7.5.1 Satisfiability (Consistency)

A sentence, $\square$, of SL is *satisfied* by a **TV** assignment, $\mathcal{A}$, iff $\square$ is assigned the **TV** t (true) by $\mathcal{A}$.

A sentence, $\square$, of SL is *satisfiable (consistent)* iff $\square$ is satisfied by at least one **TV** assignment $\mathcal{A}$.

That is, a sentence $\square$ is *satisfiable (consistent)* iff it is assigned the value t in at least one row of its truth table.

To determine whether a sentence $\square$ is *satisfiable (consistent)*, we have a straightforward procedure:

(1) Construct a truth table for $\square$.

(2) Look to see whether in at least one row, $\square$ takes the **TV** t.

(3) If so, $\square$ is satisfiable; if not, that is, if there is no row where $\square$ is assigned t, then $\square$ is not satisfiable.

A set of sentences of SL, Γ, is *satisfiable (consistent)* iff there is at least one **TV** assignment $\mathcal{A}$ that satisfies every sentence in Γ.

To determine whether a set of sentences, $\Gamma=\{\square_1, \square_2,..., \square_n\}$ is satisfiable, our procedure is this:

(1) Construct a truth-table with each of the sentences in Γ listed

across the right-hand side, with all atomic sentences occurring in any of these sentences listing at the top of the left-hand side.

(2) Determine whether there is at least one row in which each of the sentences in Γ is assigned the **TV** t.

(3) If so, the set of sentences is satisfiable; if not, that is, if there is no row where all sentences of Γ are assigned t, then Γ is not satisfiable.

If a set of sentences, $\Gamma=\{\Box_1, \Box_2,..., \Box_n\}$ is satisfiable, there must be a row of the truth-table that looks like this:

atomic sentences listed here	Sentences of Γ listed here				
	$\Box_1$	$\Box_2$	$\Box_3$	...	$\Box_n$
This is the row to look for ☞	t	t	t	t	t

If there is no row where all the $\Box$s take the value t (in a single row), then the set is inconsistent. That is, there is no assignment (row, possibility) where all the sentences are true; that is, at least one of the sentences of the set is false in every row.

atomic sentences listed here	$\Box_1$	$\Box_2$	$\Box_3$	...	$\Box_n$
	f	t	t	t	t
	t	**f**	t	t	t
	t	t	t	**f**	t
Etc.					

7.5.2 Logical Truth (Tautology)

A sentence $\Box$ is *logically true (tautologous)* in SL iff every TV assignment $\mathcal{A}$ satisfies $\Box$. That is, a sentence $\Box$ is *logically true (tautologous)* iff every row of its truth-table assigns it the value t: a logically true sentence is one that has a column of ts under it in its truth-table.

We have a *decision procedure[1]* to determine whether a sentence $\Box$ of SL is logically true:

(1) Construct a truth-table for $\Box$.

(2) Determine whether the column under $\Box$ is a column of ts; that is, every **TV** assignment assigns $\Box$ the **TV** t.

(3) If so, then □ is logically true; if not, that is, if there is even one f in the column under □, then □ is not logically true.

Graphically:

atomic sentences listed here	□
	t
	t
	t
	t
	t
	t
	t

☞ This is the definitive column; if it has all ts, the sentence is logically true.

A paradigm of a logically true sentence is any sentence of the form "□ V~□"; another is any sentence of the form "~(□ & ~□)."

7.5.3 Logical Falsehood (Self-Contradiction)

A sentence, □, is *logically false* (*self-contradictory*) in SL iff no TV assignment *A* satisfies □.

That is, a sentence □ is *logically false* (*self-contradictory*) iff every row of its truthtable assigns □ the value f; that is, iff □ has a column of fs under it in its truth-table.

We have a decision procedure to determine whether a sentence, □, is logically false:

(1) Construct a truthtable for □.

(2) Inspect the column under □ to determine whether it is a column of fs; that is, whether every TV assignment assigns □ the TV f.

(3) If so, □ is logically false; if not, if there is even one assignment of t to □, then □ is not logically false, it is satisfiable.

68

Graphically:

atomic sentences listed here	$\Box$
	f
	f
	f
	f
	f

☞ This is the definitive column; if it has all fs, the sentence is logically false.

A paradigm of a logical falsehood is any sentence of the form "$\Box$ & ~$\Box$."

7.5.4 Contingent Statements

A sentence $\Box$ is *contingent* (*logically indeterminate*) iff $\Box$ is neither logically true nor logically false.

That is, a sentence, $\Box$, is *contingent* iff its truth table shows it is true in at least one row (so it is not logically false) and that it is false in at least one row (so it is not logically true). The column in a truth table under a contingent sentence will contain at least one occurrence of a t and at least one occurrence of an f.

7.5.5 Logically Equivalent Sentences

A pair of sentences, $\Box$, Δ is *logically equivalent* iff every **TV** assignment, $\mathcal{A}$, assigns $\Box$ and Δ the same truth value.

That is, a pair of sentences, $\Box$, Δ is *logically equivalent* iff in every row of their truth table the **TV** assigned to $\Box$ is the same as the **TV** assigned to Δ.

We have a decision procedure for determining when two sentences $\Box$, Δ are logically equivalent:

(1) Construct a truthtable with $\Box$, Δ on the right-hand side.

(2) Compare the column under $\Box$ with the column under Δ to determine whether the two columns match exactly,

row for row; that is, determine whether each **TV** assignment (each row) assigns exactly the same **TV** to □ and to Δ.

(3) If the columns match row for row, then □ and Δ are logically equivalent; if not, if in even one row the sentences are assigned different **TV**s, then the sentences are not logically equivalent.

7.5.6 Contradictory Sentences

A pair of sentences □, Δ is *contradictory* iff they have different **TV**s on every **TV** assignment,𝒜. That is, a pair of sentences □, Δ is *contradictory* iff in every row of their truth table they have opposite **TV**s.

7.5.7 Logical Consequences

A sentence of SL Δ is a *logical consequence* of a set of sentences, Γ iff every **TV** assignment 𝒜 that satisfies all the sentences in Γ also satisfies Δ; that is, there is no **TV** assignment, 𝒜, that satisfies all the sentences of Γ but fails to satisfy Δ.

That is, if a sentence Δ is a logical consequence of a set of sentences Γ, no row of a truthtable assigns the **TV** t to each member of Γ but assigns the **TV** f to Δ.

We have a decision procedure for determining whether a sentence Δ is a logical consequence of a set of sentences Γ= {□₁, □₂, ..., □ₙ}:

(1) Construct a truthtable with $\square_1$, $\square_2$, ..., $\square_n$, across the right-hand side.

(2) Determine whether there is any row where each of $\square_1$, $\square_2$, ..., $\square_n$ is assigned the **TV** t and where Δ is assigned the **TV** f.

(3) If there is, Δ is not a logical consequence of Γ; if there is not, Δ is a logical consequence of Γ.

Graphically construct a truthtable as directed and look for the following sort of row:

70

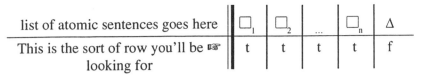

list of atomic sentences goes here	$\Box_1$	$\Box_2$	...	$\Box_n$	Δ
This is the sort of row you'll be ☞ looking for	t	t	t	t	f

If you find a row like this, then you know that Δ is not a logical consequence of Γ.

7.5.8 Truth Tables For Arguments

An argument with premises $\Box_1$, $\Box_2$, ..., $\Box_n$ and conclusion Δ is *valid* iff every **TV** assignment $\mathcal{A}$ that satisfies all the premises $\Box_1$, $\Box_2$, ..., $\Box_n$ also satisfies the conclusion Δ; that is, no **TV** assignment $\mathcal{A}$ satisfies all the premises $\Box_1$, $\Box_2$, ..., $\Box_n$ but does not satisfy the conclusion Δ.

That is, there is no row in the truth table with the sentences of the argument on its right-hand side where each of the premises of the argument have the **TV** t but the conclusion of the argument has the **TV** f.

This provides us with a decision procedure for determining whether an argument is valid:

(1) Construct a truth table, listing all the premises and the conclusion of the argument at the top of the right-hand side of the truth table, and fill out all the rows of all the columns of the truth table.

(2) Examine the column under the conclusion. If there are no fs in that column, the argument is valid. (This is because if it is impossible for a conclusion to be false, then it is impossible for it to be false while all the premises are true. So the argument is valid.) If an f occurs in the column under the conclusion, look across that row in which the f occurs to the columns under each premise. If each premise is t in that row, the argument is not valid; if one or more premise is f, look for another row where there is an f under the conclusion.

(3) If there is no row in the truth table where all the premises are true but the conclusion is false, the argument is valid. If there are one or more rows where each premise

71

is true in that row but the conclusion is false in that row, the argument is invalid.

Graphically:

Let the premises of the argument be $\square_1$, $\square_2$, $\square_3$, ..., $\square_n$, and let the conclusion be Δ. Then

list of atomic sentences goes here	$\square_1$	$\square_2$	$\square_3$	...	$\square_n$	Δ
	t	f	t	t	t	f
Counterexample! This is the ☞ sort of row you're looking for to show invalid.	t	t	t	t	t	f
	t	t	t	t	t	t

SUMMARY

Semantical Properties and Truth Tables

(1) Validity

An *argument* ARG:

$\square_1$

$\square_2$

....

$\square_n$ / $\therefore \Delta$

is *valid* iff no **TV** assignment makes its conclusion, Δ, false but satisfies all its premises, $\square_1$, ..., $\square_n$.

A truth table for a valid argument ARG will <u>not</u> have a row (i.e., a possibility) where each premise, $\square_1$, $\square_2$, ..., $\square_n$, is true and the conclusion, Δ, is false. Such a row would be a <u>counterexample</u>. So ARG is valid provided no row in a truth table for ARG is a counterexample.

(2) A sentence, $\square$, is *logically true* (a tautology) iff <u>every</u> **TV** assignment satisfies $\square$.

A truth table for a logically true $\square$ will not have a row (i.e., a possibility) where $\square$ is false. I.e., the column under $\square$ will be a column of ts.

72

(3) A sentence, $\square$, is *logically false* iff no **TV** assignment satisfies $\square$.

A truth table for a logically false $\square$ will not have a row (i.e., a possibility) where $\square$ is true. I.e., the column under $\square$ will be a column of fs.

(4) A sentence, $\square$, is *contingent* iff $\square$ is not logically true and $\square$ is not logically false.

A truth table for a contingent $\square$ will show at least one row (possibility) where $\square$ is t and at least one row where $\square$ is f. I.e., the column under $\square$ will consist of at least one t and at least one f.

(5) Two sentences, $\square$, Δ are *logically equivalent* iff every **TV** assignment that satisfies $\square$ also satisfies Δ and every **TV** assignment that does not satisfy $\square$ also fails to satisfy Δ.

A truth table for logically equivalent sentences $\square$, Δ will have no row (no possibility) where the **TV** for $\square$ differs from the **TV** of Δ. I.e., the column under $\square$ will match the column under Δ row for row.

(6) A set of sentences, $\square_1$, $\square_2$, ..., $\square_n$ is *consistent* iff it is satisfiable.

A truth table for a consistent set of sentences $\square_1$, $\square_2$, ..., $\square_n$ will have at least one row where each of the sentences of the set, $\square_1$, $\square_2$, ..., $\square_n$ is t.

(7) A set of sentences $\square_1$, $\square_2$, ..., $\square_n$ is *inconsistent* iff it is not satisfiable.

A truth table for an inconsistent set of sentences $\square_1$, $\square_2$, ..., $\square_n$ will have no row (possibility) where each of the sentences is t in that row.

7.6 Indirect (Short-Cut) Truth Tables

Semantic properties can be tested more efficiently by constructing indirect truth tables. In this method, one attempts to construct a counterexample (CEG). This is done by assigning certain **TV**s to the sentences that occur in the problem that constitute a CEG (that is, to the sentences that would occur on the <u>right-hand side</u> of the full truth table). On the basis of this assignment, **TVs** are assigned to constituents of those sentences; this process is continued until either all atomic sentences have been assigned **TVs** or until some sentence must be assigned both **TVs**.

The hypothesis that an argument is valid is correct iff there is no CEG to it. A CEG (to the hypothesis that the argument is valid) is a row of a truth table where each and every premise of the argument is true and where the conclusion is false. (That would be a row where each of the premises is true and where the negation of the conclusion is true.)

The hypothesis that a sentence is logically true is correct iff there is no CEG to it. A CEG (to that hypothesis) is a row of a truth table where the sentence takes the value "false." (That would be a row where the negation of the sentence takes the value "true".)

The hypothesis that a sentence is logically false (a self-contradiction or contradiction) is correct iff there is no CEG to it. A CEG (to that hypothesis) is a row of a truth table where the sentence takes the value "true."

The hypothesis that two sentences, □, ○, are logically equivalent is correct iff there is no CEG to that hypothesis. A CEG (to that hypothesis) is a row of a truth table where the truth values of □ and of ○ in that row differ. (That would be a row where the biconditional □ <=> ○ is "false," and therefore a row demonstrating that the biconditional is not logically true.)

The hypothesis that a set of sentences is inconsistent is correct iff there is no CEG to it. A CEG (to that hypothesis) is a row of a truth table where each of the sentences of the set has the value "true" in that row. A CEG to the hypothesis would be a truth table where there are one or more rows in which each sentence of the set has the value "true."

The hypothesis that a set of sentences is consistent is correct iff that set of sentences is not inconsistent.

CHAPTER 8

Sentence Logic: Truth Trees

8.1 Truth Tree Rules for Sentential Connectives

Truth trees are a graphical way of applying the method of counterexamples to problems.

Call any sentence that is either an atomic sentence or the negation of an atomic sentence a *literal*. For every sentence of SL that is not a literal, a decomposition tree rule will apply to that sentence. Tree rules do not apply to literals because they cannot be further decomposed. There are eight kinds of SL sentences besides literals: conjunctions, disjunctions, conditionals and biconditionals, and negations of each of these. Consequently, we have eight tree rules. It is useful to adopt a rule for double negatives as well. The result of applying a rule is written at the bottom of every open branch on which the original sentence occurs. After a rule is applied to a sentence in a tree, the sentence is checked. Finally, a rule for closing branches of a tree is required.

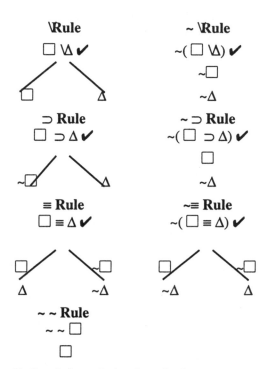

Closure Rule: A branch is closed when a sentence, □, and its negation, ~□, occurs on that branch. Branches are closed by placing an "X" at the bottom of the branch.

8.2 Semantic Properties as Characterized by Truth Trees

Semantic Property of SL	Initial Sentences of Truth Trees	Results	
		All Paths Close (i.e., the tree is closed)	One or More Paths Remain Open
□ is logically true	~□	□ is logically true	□ is not logically true
□ is logically false	□	□ is logically false	□ is not logically false
□ is contingent	Do tests for logical truth and logical falsity: if □ is neither logically true nor logically false, then □ is contingent.		
□, Δ are logically equivalent	~(□ ≡ Δ)	□, Δ are logically equivalent	□, Δ are not logically equivalent
□ is consistent / inconsistent	□	□ is inconsistent (unsatisfiable)	□ is consistent (satisfiable)
$\{\square_1, \square_2, ..., \square_n\}$ is consistent / inconsistent	$\square_1$ $\square_2$ ... $\square_n$	$\{\square_1, \square_2, ..., \square_n\}$ is inconsistent (unsatisfiable)	$\{\square_1, \square_2, ..., \square_n\}$ is consistent (satisfiable)
$\{\square_1, \square_2, ..., \square_n\}$ $\models \Delta$	$\square_1$ $\square_2$ ... $\square_n$ ~Δ	$\{\square_1, \square_2, ..., \square_n\} = \Delta$	$\{\square_1, \square_2, ..., \square_n\}$ $\nvDash \Delta$
The argument $\square_1$ $\square_2$ ..., $\square_n$ / ∴ Δ is valid	$\square_1$ $\square_2$ ... $\square_n$ ~Δ	The argument is valid	The argument is invalid

A *completed open branch* of a truth tree is a branch on which every sentence is either a literal or has been checked (has been decomposed by the application of a tree rule to it).

A *completed tree* is a tree each branch of which is either closed or is a completed open branch.

An *open tree* is one that has at least one completed open branch.

A *closed tree* is one that has no open branches.

An open branch of a truth tree corresponds to a row in a truth table where **TV**s are assigned to literals as follows: each atomic sentence that occurs as a full line on that open branch is assigned the **TV** t and each negated atomic sentence appearing as a full line on the open branch is assigned the **TV** f. Each such **TV** assignment will make each of the initial sentences of the tree true.

CHAPTER 9

Sentence Logic: Derivations

9.1 Sentence Logic: Derivations

There is considerable variation in detail among various logic texts, mainly in the rules of inference that are adopted. Some alternative rules are described in section 9.2.

The concept of a *derivation* is understood relative to a set of rules of inference and a system for constructing derivations, SD. A consecutively numbered sequence of sentences of SL is a *derivation in SD* of a sentence Δ from a set of sentences Γ of SL iff

(1) the sequence is a finite but non-empty sequence of sentences;

(2) the last sentence of the sequence is Δ;

(3) each sentence of the sequence is either (i) a member of the set Γ, or (ii) an assumption with an indication of its scope by a vertical line extending until the assumption is discharged, or (iii) an axiom of SD, or (iv) the result of applying a rule of inference of SD to one or more sentences preceding it in the sequence.

Various derivation systems may not countenance assumptions or may make no provision for a set of axioms, in which case the appropriate clause in (3) is void.

The system SD described below is a typical "Fitch-style" natural deduction system. No axioms are set out in such a system. Premises

79

(if any) are treated as primary assumptions (members of the set Γ), placed at the beginning of the derivation, and their scope is the entire derivation. Auxiliary assumptions are permitted; each auxiliary assumption is placed at the beginning of a subderivation which begins a new scope line. All assumptions are marked off by a short horizontal line intersecting with the vertical scope line of the derivation or subderivation with which the assumption is associated. Auxiliary assumptions and their scope lines are indented.

9.2 Inference Rules For Sentence Logic: Introduction and Elimination Rules

There is an introduction and an elimination rule for each connective, as well as a rule permitting reiteration of a sentence from an earlier line in a derivation to a later line under certain conditions. Many of the rules have commonly used names besides their "Int" and "Elim" names; these are given in parentheses following the "IntElim" names.

Reiteration: Reit (R)

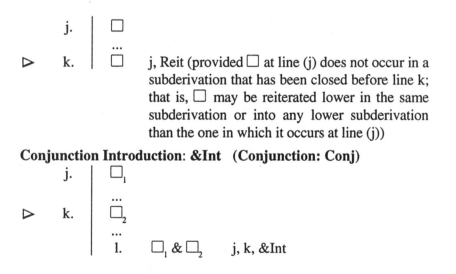

j.   □

...

▷ k.   □   j, Reit (provided □ at line (j) does not occur in a subderivation that has been closed before line k; that is, □ may be reiterated lower in the same subderivation or into any lower subderivation than the one in which it occurs at line (j))

Conjunction Introduction: &Int   (Conjunction: Conj)

j.   □₁

...

▷ k.   □₂

...

l.   □₁ & □₂   j, k, &Int

Conjunction Elimination: &Elim (Simplification: Simp)

$$j. \quad \Box_1 \And ... \And \Box_i \And ... \And \Box_n$$

$$\triangleright \quad k. \quad \Box_i \qquad j, \&Elim$$

Disjunction Introduction: VInt (Addition: Add)

$$j. \quad \Box_i$$

$$\triangleright \quad k. \quad \Box_1 \lor ... \lor \Box_i \lor ... \Box_n \qquad j, VInt$$

Disjunction Elimination: VElim (Argument by Cases)

$$j. \quad \Box_1 \lor ... \lor \Box_n$$

$$k_1. \quad \vdash \quad \Box_1$$
$$k_n. \quad \Delta$$

etc. for each $\Box_i$ in the disjunction on line (j)

$$l_1. \quad \vdash \quad \Box_n$$
$$l_n. \quad \Delta$$
$$m. \quad \Delta \qquad j, k_1 - k_n, l_1 - l_n, VElim$$

Negation Introduction: ~Int (Reductio ad Absurdum: RAA; Indirect Proof: IP)

$$j. \quad \vdash \quad \Box$$
$$k. \quad \Delta \And {\sim}\Delta$$
$$\triangleright \quad l. \quad {\sim}\Box \qquad j - k, {\sim}Int$$

Negation Elimination: ~Elim (Reduction ad Absurdum: RAA; Indirect Proof: IP)

$$j. \quad - \quad {\sim}\Box$$
$$k. \quad \Delta \And {\sim}\Delta$$
$$\triangleright \quad l. \quad \Box \qquad j - k, {\sim}Elim$$

Conditional Introduction: Int (Conditional Proof: CP)

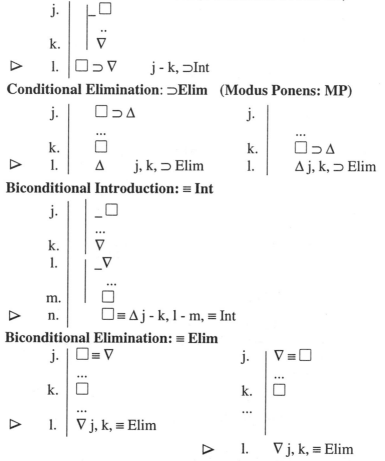

j.   ⌐□

k.   ∇

▷ l. □ ⊃ ∇   j - k, ⊃Int

Conditional Elimination: ⊃Elim (Modus Ponens: MP)

j.   □ ⊃ Δ   j.

k.   □   k.   □ ⊃ Δ

▷ l. Δ   j, k, ⊃ Elim   l.   Δ j, k, ⊃ Elim

Biconditional Introduction: ≡ Int

j.   _□

k.   ∇

l.   _∇

m.   □

▷ n. □ ≡ Δ j - k, l - m, ≡ Int

Biconditional Elimination: ≡ Elim

j. □ ≡ ∇   j. ∇ ≡ □

k. □   k. □

▷ l. ∇ j, k, ≡ Elim

                        ▷ l. ∇ j, k, ≡ Elim

9.3 Additional Derivation Rules

The following rules can be derived in the system SD: anything that can be derived in SD using them can be derived in SD without using them. They allow for shorter and easier derivations in many instances. One kind of derived rule works like the IntElim rules above: they apply to whole lines of derivations. Another kind of rule may apply to whole lines or to parts of sentences occurring on lines of derivations; these are called "Rules of Replacement."

9.3.1 Derived Inference Rules

Modus tollens: MT

j.	$\Box \supset \Delta$j.			~Δ	
	...			...	
k.	~Δ		k.	$\Box \supset \Delta$	
▷ l.	~$\Box$	j, k, MT	l.	~$\Box$	j, k, MT

Disjunctive Syllogism: DS

j.	$\Box \vee \Delta$		j.	$\Box \vee \Delta$	
	...			...	
k.	~$\Box$		k.	~Δ	
	...			...	
▷ l.	Δ	j, k, DS	▷ l.	$\Box$	j, k, DS

Hypothetical Syllogism: HS (Transitivity of $\supset$: Trans $\supset$)

j.	$\Box \supset \Delta$	
k.	$\Box \supset \bigcirc$	
▷ l.	$\Box \supset \bigcirc$	j, k, HS

Transitivity of $\equiv$: Trans $\equiv$

j.	$\Box \equiv \Delta$	
	...	
k.	$\Delta \equiv \bigcirc$	
	...	
▷ l.	$\Box \equiv \bigcirc$	j, k, Trans $\equiv$

Constructive Dilemma: CD

j.	$\Box \vee \Delta$	
	...	
k.	$\Box \supset \bigcirc$	The sentences on lines (j), (k), and (l) may occur in any order.
	...	
l.	$\Box \supset \Diamond$	
	...	
▷ m.	$\bigcirc \vee \Diamond$	j, k, l, CD

83

Destructive Dilemma: DD

j.	~○ V ~ ◊
	...
k.	□ ⊃ ○ The sentences on lines (j), (k), and (l) may occur in any order.
	...
l.	Δ ⊃ ◊
	...
m.	~ □ V ~ Δ j, k, l, DD

Strengthening the Antecedent: (Str Ant)

	j.	□ ⊃ ∇
		...
▷	k.	(○ & □) ⊃ ∇ j, StrAnt

Weakening the Consequent: (Wk Con)

	j.	□ ⊃ Δ
		...
▷	k.	□ ⊃ (Δ V○) j, WkCon

9.3.2 Replacement Rules

Replacement rules may apply to whole sentences or parts of sentences. This is because they formulate logical equivalences. Let "□ :: Δ" represent the fact that either sentence may replace the other at one or more of its occurrences.

Generalized Substitution: (Gen Sub)

j.	□ ≡ Δ
	...
k.	_ _ _ □ _ _ _ where "_ _ _" represents any sentential context
	...
l.	_ _ _ Δ _ _ _ j, k GenSub

Double Negation: DN

_ _ _ □ _ _ _ :: _ _ _ ~ ~ □ _ _ _

Idempotence of &: Idem & (Tautology &: Taut &)

_ _ _ [□ & □] _ _ _ :: _ _ _ □ _ _ _

84

Idempotence of V: Idem V(Tautology &:Taut &)

$\_\_\_[\Box\ V\Box]\_\_\_::\_\_\_\Box\_\_\_$

Commutativity of &: Com&

$\_\_\_[\Box\ \&\ \Delta]\_\_\_::\_\_\_[\Delta\ \&\ \Box]\_\_\_$

Commutativity of V: Com V

$\_\_\_[\Box\ V\Delta]\_\_\_::\_\_\_[\Delta\ V\Box]\_\_\_$

Commutation of ≡: Com ≡

$\_\_\_[\Box\equiv\Delta]\_\_\_::\_\_\_[\Delta\equiv\Box]\_\_\_$

Association of &: Assoc &

$\_\_\_[(\Box\ \&\ \Delta)\ \&\ \bigcirc]\_\_\_::\_\_\_[\Box\ \&\ (\Delta\ \&\ \bigcirc)]\_\_\_$

Association of V: Assoc V

$\_\_\_[(\Box\ V\Delta)\ V\bigcirc]\_\_\_::\_\_\_[\Box\ V(\Delta\ V\bigcirc)\ ]\_\_\_$

Association of ≡: Assoc ≡

$\_\_\_[(\Box\equiv\Delta)\equiv\bigcirc]\_\_\_::\_\_\_[\Box\equiv(\Delta\equiv\bigcirc)\ ]\_\_\_$

DeMorgan's Laws: (DeM)

$\_\_\_[\sim(\Box\ \&\ \Delta)]\_\_\_::\_\_\_[\sim\!\Box\ V\!\sim\Delta]\_\_\_$

$\_\_\_[\sim(\Box\ V\Delta)]\_\_\_::\_\_\_[\sim\!\Box\ \&\sim\Delta]\_\_\_$

DeMorgan's Laws apply to conjunctions, disjunctions, negations of conjunctions, and negations of disjunctions. DeMorgan's Laws can be applied by performing the following steps:

(1) Negate the entire molecular sentence (conjunction, disjunction, negation of conjunction, or negation of disjunction);

(2) Negate each immediate constituent;

(3) Change "&" to "V" or "V" to "&," as appropriate.

Distribution of & over V: (Dist&)

$\_\_\_[\Box\ \&\ (\Delta\ V\bigcirc)]\_\_\_::\_\_\_[(\Box\ \&\ \Delta)\ V(\Box\ \&\ \bigcirc)]\_\_\_$

Distribution of V over &: (Dist V)

$\_\_\_[\Box\ V\ (\Delta\ \&\bigcirc)]\_\_\_::\_\_\_[(\Box\ V\Delta)\ \&(\Box\ V\bigcirc)]\_\_\_$

Contraposition: Contrap (Transposition: Trans)

$\_\_\_ [\square \supset \triangle] \_\_\_ :: \_\_\_ [\sim\triangle \supset \sim\square] \_\_\_$

Implication: Impl (Material Implication: Matl Impl)

(i) $\_\_\_ [\square \supset \triangle] \_\_\_ :: \_\_\_ [\sim\square \lor \triangle] \_\_\_$

(ii) $\_\_\_ [\sim\square \supset \triangle] \_\_\_ :: \_\_\_ [\square \lor \triangle] \_\_\_$

(iii) $\_\_\_ [\sim (\square \supset \triangle)] \_\_\_ :: \_\_\_ [\square \,\&\, \sim\triangle] \_\_\_$

Importation/Exportation (Imp/Exp)

$\_\_\_ [\square \supset (\triangle \supset \bigcirc)] \_\_\_ :: \_\_\_ [(\square \,\&\, \triangle) \supset \bigcirc] \_\_\_$

Equivalence (Equiv)

(i) $\_\_\_ [\square \equiv \triangle] \_\_\_ :: \_\_\_ [(\square \supset \triangle) \,\&\, (\triangle \supset \square)] \_\_\_$

(ii) $\_\_\_ [\square \equiv \triangle] \_\_\_ :: \_\_\_ [(\square \,\&\, \triangle) (\sim\square \,\&\, \sim\triangle)] \_\_\_$

9.4 Basic Concepts of Derivation Systems for Sentence Logic

A sentence, $\triangle$, is *derivable in SD* from a set of sentences, Γ, ($\Gamma \vdash_{SD} \triangle$) iff there is a derivation in SD of $\triangle$ from Γ.

A sentence, $\triangle$, is a *theorem* ($\vdash_{SD} \triangle$) of SD iff $\triangle$ is derivable from the empty set of premises.

Sentences $\square$, $\triangle$ are *equivalent* in SD iff $\{\square\} \vdash_{SD} \triangle$ and $\{\triangle\} \vdash_{SD} \square$ (that is, two sentences are equivalent in SD iff they are derivable from each other).

An argument of SL,

$$\square_1, \square_2, ..., \square_n \,/\, \therefore \triangle$$

is *valid in SD* iff $\{\square_1, \square_2, ..., \square_n\} \vdash_{SD} \triangle$ (that is, iff its conclusion is derivable from its premises).

A set of sentences, Γ, is *inconsistent in SD* iff there is some sentence, $\triangle$, of SL such that both $\Gamma \vdash_{SD} \triangle$ and $\Gamma \vdash_{SD} \sim \triangle$ (that is, a set of sentences Γ is inconsistent in SD iff some sentence and its negation are both derivable from Γ).

9.5 SD Theorems

Selected theorems of SD follow. The number of theorems is (denumerably) infinite.

1. $(A \supset B) \supset [(B \supset C) \supset (A \supset C)]$
2. $[A \supset (B \supset C)] \supset [(A \supset B) \supset (A \supset C)]$
3. $A \supset [(A \supset B) \supset B]$
4. $A \supset A$
5. $A \supset (B \supset A)$
6. $A \supset (\sim A \supset B)$
7. $\sim A \supset (A \supset B)$
8. $\sim \sim A \supset A$
9. $A \supset \sim \sim A$
10. $(\sim A \supset \sim B) \supset (B \supset A)$
11. $(A \supset B) \supset (\sim B \supset \sim A)$
12. $(A \supset \sim A) \supset \sim A$
13. $\sim (A \supset B) \supset A$
14. $\sim (A \supset B) \supset \sim B$
15. $A \supset [B \supset (A \,\&\, B)]$
16. $(A \supset B) \supset [(B \supset A) \supset (A \equiv B)]$
17. $(A \equiv B) \supset (A \supset B)$
18. $(A \equiv B) \supset (B \supset A)$
19. $A \equiv A$
20. $\sim \sim A \equiv A$
21. $A \lor \sim A$
22. $\sim (A \,\&\, \sim A)$
23. $(A \equiv B) \supset [(A \,\&\, B) \lor (\sim A \,\&\, \sim B)]$
24. $\sim (A \equiv B) \equiv (A \equiv \sim B)$
25. $[(A \equiv B) \equiv A] \equiv B$
26. $(A \supset B) \equiv (\sim A \lor B)$

27. $(A \supset B) \equiv \sim (A \& \sim B)$

28. $(A \supset B) \equiv [A \equiv (A \& B)]$

29. $(A \supset B) \equiv [(A \lor B) \equiv B]$

30. $[A \supset (B \& C)] \equiv [(A \supset B) \& (A \supset C)]$

31. $[(A \lor B) \supset C] \equiv [(A \supset C) \& (B \supset C)]$

32. $[A \supset (B \lor C)] \equiv [(A \supset B) \lor (A \supset C)]$

33. $[(A \& B) \supset C] \equiv [(A \supset C) \lor (B \supset C)]$

34. $(A \supset B) \lor (B \supset A)$

35. $(A \supset B) \lor (\sim A \supset B)$

36. $(A \supset B) \lor (A \supset \sim B)$

37. $[(A \& B) \supset C] \equiv [(A \& \sim C) \supset \sim B]$

38. $[([A \supset B] \supset C) \supset D] \supset [(B \supset C) \supset (A \supset D)]$

39. $[(A \supset B) \supset A] \equiv A$

40. $[(A \supset B) \supset C] \supset [(A \supset C) \supset C]$

CHAPTER 10

Predicate Logic: Symbolization and Syntax

10.1 Vocabulary of PL

For predicate logic, the language of SL is extended to include symbols for proper names; noun, verb, adjective, and prepositional phrases; the quantifiers "all" and "some"("there exists at least one"); and identity. The symbols of SL are retained, with modest changes in the grammar of the connectives. The result is a language in which any sentence can be represented that can be represented in SL and in categorical propositions, and much more. We also develop a logic powerful enough to treat satisfactorily all that has been presented earlier and more. The language for predicate logic is called "PL."

The symbols used vary from one text to another, mostly in minor ways. Singular terms, representing proper names, are lower case letters from "a" to "t," with or without numerical subscripts (to ensure that a sufficient number are available). Predicates, representing noun, verb, adjective, and prepositional phrases, are represented by upper case letters with or without numerical subscripts and with or without numerical superscripts. Subscripts on predicate letters ensure that a sufficient number are available; superscripts indicate the number of terms (called arguments) that are required to turn the predicate into a sentence, but they are almost always omitted, since context indicates the number of arguments required. Sentence letters are 0-place predicates.

Sentence letters: "A" to "Z," with or without numerical subscripts.

Predicates: "A" to "Z," with or without numerical subscripts, and with one or more numerical superscripts, indicating the number of arguments the predicate requires to form a well-formed formula (wff).

Individual terms

Individual constants: "a" to "v," with or without numerical subscripts.

Individual variables: "w" to "z," with or without numerical subscripts.

Logical operators

Sentential connectives: $\sim$, &, V, $\equiv$,

Quantifier symbols: $\forall$, $\exists$

Two place identity predicate: =

Punctuation: (,) (which may also be written as [,]; {, }; etc.)

For variation in symbols for the connectives see Chapter 6.1.1. Minor variation occurs in the letters selected for predicates, individual constants, and individual variables. Identity may be represented by "I^2."

Quantifier symbols Let "α" represent any variable of PL.

Universal quantifier: $\forall$, (α), Λ, Π

Existential quantifier: $\exists$, V, Σ

10.2 Informal Description of PL

In PL, quantifiers occur with variables; together, they establish the kind of linkages that are found with anaphoric pronouns in natural languages, as in "All politicians think that they can fool the voters, but they can't" (the antecedent is "all politicians"). This increases the expressive power of PL over categorical propositions.

In translating English sentences, we write a *symbolization key*. This is comparable to a list of abbreviations in SL, but since there is so much more structure to deal with in PL, it is more useful to have a key to avoid confusion. In addition to indicating what letters will be used to symbolize which English names and predicates, a symbolization key must also specify a *Domain* or *Universe of Discourse* (UD).

A UD can be any non-empty set of things; the UD for a given symbolization key is (roughly) the set of all the things talked about in the particular context we're concerned with in dealing with a problem. Often the UD is not explicitly specified; it "is understood." But if it is, it must be specifiable. The UD is crucial in understanding the quantity words of PL.

The universal quantifier is understood as expressing the idea that each and every element in the UD has the property expressed by the (possibly complex) predicate following it. The existential quantifier is understood as expressing the idea that at least one member of the UD has the property expressed by the predicate following it.

Some simple translations are provided of singular sentences, of each form of categorical proposition, and of some more complex sentences.

Sample translations

Symbolization key:

Universe of Discourse: people and novels

A_	_ is an author	a 1-place predicate
H_	_ is a human being	a 1-place predicate
N_	_ is a novel	a 1-place predicate
W _ _	_ wrote _	a 2-place predicate
c	Samuel Clemens	an individual constant
t	Mark Twain	an individual constant
s	The novel Tom Sawyer	an individual constant

English sentence	Translation in PL
1. Mark Twain is an author.	At
2. Mark Twain wrote *Tom Sawyer*.	Wts
3. Mark Twain is Samuel Clemens.	$t = c$
4. All authors are human. (For every x, if x is an author, then x is human)	$(\forall x)(Ax \supset Hx)$
5. No novel is a human being. (For every x, if x is a novel, then x is not human; there doesn't exist an object which is a novel and is a human being.)	$(\forall x)(Nx \supset {\sim}Hx)$ ${\sim}(\exists x)(Nx \,\&\, Hx)$

91

6. Some human being is an author. $(\exists y)(Hy \ \& \ Ay)$

7. Some human is not an author. $(\exists z)(Hz \ \& \sim Az)$

8. Some author wrote *Tom Sawyer*. $(\exists x)(Ax \ \& \ Wxs)$

9. Everyone wrote some novel or other. $(\forall x)[Hx \supset (\exists y)(Ny \ \& \ Wxy]$

10. If Twain wrote *Tom Sawyer* then a human wrote it. $Wts \supset (\exists x)Wxs$

Translating categorical propositions into PL

Categorical Proposition	Translation in PL
A: All S are P	$(\forall x)(Sx \supset Px)$
E: No S are P	$(\forall x)(Sx \supset \sim Px)$
I: Some S are P	$(\exists x)(Sx \ \& \ Px)$
O: Some S are not P	$(\exists x)(Sx \ \& \sim Px)$

Note: A-type statements are translated as universal conditionals, and I-type statements are translated as existential conjunctions. Translating an I-type statement as "$(\exists x)(Sx \supset Px)$" is a common error. This is equivalent to the very weak "$(\exists x)(\sim Sx \lor Px)$," which says "There is something that either is not an S or is a P." The Modern Square of Opposition has E-type statements and I-type statements as contradictories; consequently, an alternative translation of an E-type statement is simply as the negation of the I-type statement.

See Chapter 3.10 for translation hints for other forms of English sentences such as temporal and spatial adverbs, nonstandard quantifiers, unexpressed quantifiers, exclusive propositions, and exceptive propositions: paraphrase these into categorical propositions, and then translate the categorical proposition(s) as indicated above.

Besides expressing quantity, modern quantifier notation also exhibits linkage. Thus in sentence (9) above, the notation shows that the universal quantifier is linked to the term denoting the human and to the object that does the writing, while the existential quantifier is linked to the term denoting novels and to the object that is written. Graphically:

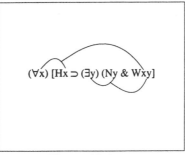

Figure 10-1

10.3 Formal Grammar of PL

10.3.1 Preliminaries: Notation

The following metavariables range over expressions of PL.

□, ○, Δ, ◊.

"ß" is a metavariable ranging over individual constants of PL and "α" is a metavariable ranging over variables of PL. "τ" is a metavariable ranging over terms of PL (i.e., expressions which may be either constants or variables). These symbols may be used with subscripts so as to have enough. None of these metavariables are in the vocabulary items of PL; they supplement the English we use to describe PL.

10.3.2 Grammar: Syntax and Definitions

(Note: the terminology varies from one text to another, as does the notation. Many texts treat quantification theory purely informally and fail to introduce many of the concepts (characterized formally) that follow in this section.)

An <u>expression</u> of PL is any sequence of elements from the vocabulary of PL.

A <u>quantifier</u> is a universal or an existential quantifier.

A <u>universal quantifier</u> is an expression of the form "(∀α)" — this means that a universal quantifier is an expression consisting of a left parenthesis followed by a "∀" followed by a variable, followed by a right parenthesis.

An <u>existential quantifier</u> is an expression of the form "($\exists\alpha$)".

A quantifier <u>contains</u> the variable occurring in it.

<u>Atomic well-formed formulas (atomic wffs)</u>:

(1) Sentence letters are atomic wffs.

(2) An n-place predicate followed by n individual terms is an atomic wff.

(3) The identity predicate flanked by individual terms is an atomic wff; i.e., any sentence of the form $\tau_1 = \tau_2$ is an atomic wff.

(4) Nothing else is an atomic wff.

<u>Atomic sentences</u>

An atomic sentence is an atomic wff all of whose terms are individual constants (names).

<u>Well-formed Formulas</u> (wffs):

(1) Atomic wffs are wffs.

(2) If $\square$ is a wff, then ~$\square$ is a wff.

(3) If $\square$, Δ are wffs, then

  (a) ($\square$ & Δ) is a wff;

  (b) ($\square$ V Δ) is a wff;

  (c) ($\square \supset \Delta$) is a wff;

  (d) ($\square \equiv \Delta$) is a wff.

(4) If $\square$ is a wff, and α is a variable, and if

  (i) $\square$ contains at least one occurrence of a variable, α and

  (ii) $\square$ does not contain an occurrence of a quantifier containing α, then:

  (a) ($\forall\alpha$)$\square$ is a wff;

  (b) ($\exists\alpha$)$\square$ is a wff.

(5) Nothing else is a wff.

The exact definition of "wff" varies among texts. In particular, some texts may omit the two conditions in clause (4). Condition (i) in clause (4) blocks expressions such as "($\forall$x)Fa" from being a wff of PL, where the quantifier is not linked to anything. This jibes with its English correlate, "For every x, Alfred is funny," for example, which is not coherent. Condition (ii) blocks expressions such as "(∀x)(∀x) Lxx" from being wffs. This is sensible, since the left-most quantifier is redundant; the quantifier on the right is linked to all occurrences of "x," so the left quantifier is linked to nothing.

Subformula (subwff) of a wff, □:

We give a recursive definition of *subformula (subwff)* and *main logical operator* of a wff:

(1) If □ is atomic, □ is the only subwff of □, and □ contains no logical operators and hence no main logical operator.

(2) If □ has the form ~Δ, then "~" is the main logical operator of □ and Δ is a subwff of □.

(3) If □ has the form (Δ & ○), (Δ ∨ ○), (Δ ⊃ ○), or (Δ ≡ ○), then the binary sentential connective is the main logical operator of □, and Δ, ○ are subwffs of □.

(4) If □ has the form (∀α)Δ or (∃α)Δ, then the quantifier is the main logical operator of □, and Δ is a subwff of □.

(5) Every subwff of a subwff of □ is a subwff of □ and □ is a subwff of itself.

Δ is a proper subwff of □ iff Δ is a subwff of □ but Δ ≠ □. (That is, no wff is a proper subwff of itself, though every wff is a subwff of itself.)

Scope of a quantifier:

The *scope* of a quantifier in a wff □ is the subwff Δ of □ of which that quantifier is the main logical operator. Scope is related to linkage: A variable that occurs outside the scope of a quantifier cannot be linked to it.

Note: the scope of a quantifier is much like the scope of "~": the shortest subwff compatible with the punctuation. The difference is that for negation, the scope is the proper subwff of the wff of which the "~" is the main logical operator; whereas for the quantifiers, the scope is the subwff of which the quantifier is the main logical operator.

Wff	Scope
(∀x)Fx	(∀x)FX
~P	P
(∀x)(Fx ⊃ Gx)	(∀x)(Fx ⊃ Gx)
~(P & Q)	(P & Q)
(∀x) Fx ⊃ Gx	(∀x)Fx
~P & Q	P

Bound and free occurrences of variables

Bound (occurrence of a) variable:

An occurrence of a variable α in a wff □ is *bound* iff that occurrence of α falls within the scope of an α-quantifier.

Free (occurrence of a) variable;

An occurrence of a variable is *free* iff it is not a bound occurrence.

We can say that an occurrence of a variable α is <u>bound</u> if it occurs within the scope of a quantifier containing α; otherwise it is a <u>free</u> occurrence of the variable. Occurrences of variables as part of the quantifier are always bound.

Examples:

In "(∃y)(Fxy & Gyzx)," all occurrences of "y" are bound, both occurrences of "x" are free, and the occurrence of "z" is free.

In "(∀x)[Fax ⊃ (∃y)(Fyx & Gyz)] ≡ Fax," the first three occurrences of "x" are bound, but the fourth is free; all three occurrences of "y" are bound; and the occurrence of "z" is free. The individual constant "a" is neither bound nor free, since the concepts "bound" and "free" are defined only for variables.

Sentence of PL

A wff □ of PL is a *sentence* (or *closed wff*) of PL iff no occurrence of a variable in □ is free.

A wff □ of PL is an *open sentence* (or *open wff*) of PL iff one or more occurrences of a variable in □ are free.

So an expression □ is a sentence of PL iff:

96

(i) ☐ is a wff of PL;

(ii) all occurrences of variables in ☐, if any, are bound.

Substitution instance

Where ☐ is any wff, α is any variable, and β is any individual constant, ☐ α/β is the result of replacing all free occurrences of α in ☐ by occurrences of β. Examples:

☐ = Fx   ☐ "x"/"a" = Fa

☐ = Gy & Hyb   ☐ "x"/ "a" = Gy & Hyb

                       ☐ "y" / "a" = Ga & Hab

                       ☐ "y"/ "b" = Gb & Hbb

Substitution instance of a quantified sentence

If ☐ has the form (∀α)Δ or (∃α)Δ and if β is an individual constant, then Δα/β is a *substitution instance* of ☐. β is called the *instantiating constant*.

The procedure to form a substitution instance of a quantified wff:

(i) Drop the quantifier that is the main logical operator of the wff. This results in an open wff that has at least one free occurrence of a variable, namely, the variable linked to the quantifier that was dropped.

(ii) Pick any constant.

(iii) Replace all free occurrences of the variable of quantification (i.e., all the occurrences that were linked to the quantifier) with occurrences of that constant.

The procedure for forming an existential generalization from a given sentence ☐ of PL:

(1) Select a constant, β, occurring in the sentence ☐.

(2) Replace some (or all) occurrences of β with occurrences of any variable, α, with the following restriction: when a given occurrence of β is replaced by an occurrence of α, the occurrence of α must not become bound by a quantifier already in the wff. For example, to existentially generalize on "a" in the sentence "(∃x)Fxa," a variable other than "x" must be selected to replace "a"; if "x" is selected, it will become bound by the quantifier already

97

occurring in the sentence, since "a" is in the scope of that quantifier. So select "y" and obtain "(∃y)(∃x)Fxy").

(3) Prefix an existential quantifier on α to the wff that results from (2). The scope of this quantifier must be the entire sentence; i.e., the prefixed existential quantifier becomes the main logical operator of the resulting sentence.

The procedure for forming a universal generalization from a sentence of PL:

(1) Select a constant, β, to generalize on. Select a variable, α, that does not occur in the given sentence as the variable of generalization. If the given sentence, □, contains one or more occurrences of a quantifier whose variable of quantification is α, you must select a different variable of quantification. This is to avoid violating conditions (i) and (ii) of clause (4) of our definition of a wff.

(2) Replace <u>every</u> occurrence of β in □ by an occurrence of α. Note: this is more restrictive than the corresponding condition for existential generalization, where one or more or all occurrences of β could be replaced, at your option. Here, each and every occurrence of β must be replaced (after the condition in (1) is met). The result is that every occurrence of α that has replaced a β must be a free occurrence; the reason for the condition in (1) is to ensure that this is the case.

(3) Prefix the wff that results from step (2) by a universal quantifier whose variable of quantification is α and whose scope is all of □. That is, the newly introduced universal quantifier must be the main logical operator of the sentence that results.

10.4 Translation from English to PL

"<u>Any</u>"

"Any" is sometimes translated by a universal quantifier; in those cases, it can be paraphrased by "every" or "all."

Anyone who enjoys desserts likes Sara Lee. (∀x)(Ex ⊃ Lxs)

Everyone who enjoys desserts likes Sara Lee. $(\forall x)(Ex \supset Lxs)$

But sometimes "any" is not simply synonymous with "every":

(1) If everyone enjoys desserts, Alicia does. $(\forall x)Ex \supset Ea$

(2) If anyone enjoys desserts, Alicia does. $(\exists x)Ex \supset Ea$.

Note: these statements are conditionals. The first is trivially true, saying "if everything is a certain way, then Alicia is that way"; the second is not trivially true, saying, in effect, that if at least one thing is a certain way, then Alicia is that way. Finally compare (2) to (3):

(3) If anyone enjoys desserts, they enjoy coffee. $(\forall x)(Ex \supset Cx)$

(3) cannot be translated correctly by "$(\exists x)Ex \supset Cx$," because this has a free occurrence of "x" and so is not a sentence of PL. Note that the English (3) has an anaphoric pronoun, "they," linked to its antecedent "anyone." When an English sentence contains a quantified noun phrase in the antecedent of a conditional, and there is an (explicit or implied) anaphoric pronoun linked to that quantified noun phrase, a universal quantifier whose scope is the entire conditional is required to translate the sentence.

"Not any"

Some texts advise that "not any ..." = "none ..." = "~ $(\exists x)$ ...";
and that "not every ..." = "~ $(\forall x)$ ...." These rules are at best guidelines that must be evaluated in each case. The advice will not yield a correct translation when applied to the sentence

Not just any player can make that move. $\sim(\exists x)Mx$

which allows that Michael Jordan can make that move, but the translation states incorrectly that no one can make that move.

<u>Noun Phrases</u>

Proper names translate as individual constants. Common nouns generally translate as monadic predicates, but some English common nouns may be analyzed as relations: for example, "is a father" is a monadic predicate that can be analyzed when the context warrants as a second-place predicate "is a father of someone."

<u>Adjectives</u>

Adjectives usually modify nouns: <u>brick</u> school, <u>blue</u> bell, etc. and are translated, with a few exceptions, as monadic predicates.

Exceptions: large mouse (is a mouse but isn't large); suspected thief (is suspected but may not be a thief); short giraffe (is a giraffe but may not be short); similarly for adjectives such as heavy/light/tall/small, alleged criminal, fake diamonds, former senator, etc. The exceptional cases involve an interaction between the meaning of the adjective and the noun it modifies: large for a mouse, small for an elephant, suspected of being a criminal, etc. The exceptions can make a difference in evaluating arguments. The argument

Every philosopher is a lover./∴Every good philosopher is a good lover.

is not a valid argument, in contrast with

All buildings are structures/∴ All brick buildings are brick structures.

Translate the exceptions as single monadic predicates: "large mouse" : Mx.

Adjectives modified by adverbs such as very wealthy lawyer are not translated as a conjuction like "x is very & x is wealthy & x is a lawyer" but rather by " x is very wealthy & x is a lawyer."

Relative clauses

Relative clauses are expressions formed from sentences. They begin with words like that or with a wh- word such as who, what, which, though often these words are omitted. Often relative clauses modify nouns or noun phrases, like adjectives. Translate them as conjoined to predicates symbolizing nouns:

A car (that) I used to drive was scrapped = (∃x) (Cx & Dix & Sx).

Restrictive relative clauses are contrasted with appositive relative clauses (the latter are or could be set off by parentheses or commas). Both kinds are always translated as conjunctions. But where there's a universal quantifier and the relative clause modifies the subject noun phrase, a restrictive relative clause such as (1) is translated differently from an appositive relative clause, (2), as illustrated by the examples:

(1) All the aldermen who have been convicted are claiming to have medical problems.

$(\forall x)[(Ax \& Cx) \supset Mx]$ :: All convicted aldermen are claiming ... (not necessarily all aldermen)

100

(2) All the aldermen, who have been convicted, are claiming to have medical problems.

($\forall$x)(Ax $\supset$ Cx) & ($\forall$x)(Ax $\supset$ Mx)   All the aldermen are claiming ... and all have been convicted

(1) is a universally quantified conditional whose antecedent is a conjunction; (2) is a conjunction of universally quantified sentences; they are not logically equivalent.

Verb Phrases

Verbs fall into two general nonexclusive classes: intransitive and transitive.

Intransitive Verbs: Do not take objects; translate into monadic predicates:

Someone laughed. = ($\exists$x) (Px & Lx)

Everyone dies. = ($\forall$x) (Px $\supset$ Dx)

Transitive Verbs: Take direct objects and may take indirect objects; these translate into polyadic predicates:

x loves y = Lxy

Someone loves everyone. = ($\exists$x)($\forall$y) Lxy

x gives y to z = Gxyz

Some man gave a woman a scarf. = ($\exists$x)[Mx & ($\exists$y)(Wy & ($\exists$z) (Sz & Gxzy)]

Verbs whose objects are clausal complements take as objects sentences or grammatical constructions closely related to sentences as objects: believe, know, persuade, hope, want, etc. These create non-extensional contexts and are translated as atomic sentences.

Prepositional Phrases

Prepositions that modify nouns to form prepositional phrases commonly can be translated as second-place, third-place, etc. predicates:

Everyone from Chicago is a Bulls' fan. = ($\forall$x) [(Px & Fxc) $\supset$ Bx]

I'm between a rock and a hard place. = ($\exists$x)(Rx & ($\exists$y) (Hy & Bixy)

Quantifier Negation

Negations of quantified sentences can be translated simply by writing the quantified sentence and then prefixing a negation sign to it. Any such sentence is logically equivalent to another sentence which is not a negation, as seen by reflecting on the Modern Square of Opposition.

The general mechanism of quantifier negation is simple: moving a negation from one side of a quantifier to the other changes the quantifier to its dual (i.e., changes an existential to a universal and changes a universal to an existential).

Thus, the negation of an A-statement is an O-statement:

$\sim(\forall x)Fx$ :: $(\exists x) \sim Fx$

$\sim(\forall x)(Fx \supset Gx)$ :: $(\exists x) \sim(Fx \supset Gx)$ :: $(\exists x)(Fx \ \& \sim Gx)$

The negation of an I-statement is an E-statement:

$\sim(\exists x)Fx$ :: $(\forall x) \sim Fx$

$\sim(\exists x)(Fx \ \& \ Gx)$ :: $(\forall x) \sim (Fx \ \& \ Gx)$ :: $(\forall x)(Fx \supset \sim Gx)$

The negation of an E-statement is an I-statement:

$\sim(\forall x) \sim Fx$ :: $(\exists x)Fx$

$\sim (\forall x)(Fx \supset \sim Gx)$ :: $(\exists x) \sim(Fx \supset \sim Gx)$ :: $(\exists x)(Fx \ \& \ Gx)$

Note that if the E-statement is symbolized as $\sim(\exists \alpha) \Box \alpha$, then it is obvious that the negation of an E-statement is an I-statement (by double negation).

The negation of an O-statement is an A-statement:

$\sim(\exists x) \sim Fx$ :: $(\forall x)Fx$

$\sim(\exists x)(Fx \ \& \sim Gx)$ :: $(\forall x) \sim(Fx \ \& \sim Gx)$ :: $(\forall x)(Fx \supset Gx)$

Quantifier scope equivalences

The following quantificational equivalences involving quantifier scope changes are useful because they illustrate different but logically equivalent ways an English sentence can be symbolized. Where α does not contain an occurrence of the variable α:

(1) $(\exists \alpha) \Box \alpha \lor \bigcirc$ :: $(\exists \alpha)(\Box \alpha \lor \bigcirc)$

(2) $(\forall \alpha) \Box \alpha \lor \bigcirc$ :: $(\forall \alpha)(\Box \alpha \lor \bigcirc)$

(3)   ○ V(∃α) □α   : :   (∃α)(○ V□α)

(4)   ○ V(∀α) □α   : :   (∀α)(○ V□α)

(5)   (∃α) □α & ○   : :   (∃α)(□α & ○)

(6)   (∀α) □α & ○   : :   (∀α)(□α & ○)

(7)   ○ & (∃α) □α   : :   (∃α)(○ & □α)

(8)   ○ & (∀α) □α   : :   (∀α)(○ & □α)

When the quantifier occurs in the <u>consequent</u> of a conditional, the equivalences work as usual:

(9)   ○ ⊃ (∃α) □α   : :   (∃α)(○ ⊃ □α)

(10)   ○ ⊃ (∀α) □α   : :   (∀α)(○ ⊃ □α)

When the quantifier occurs as the <u>antecedent</u> of a conditional, however, the equivalences have a different pattern:

(11)   (∃α) □α ⊃ ○   : :   (∀α)(□α ⊃ ○)

(12)   (∀α) □α ⊃ ○   : :   (∃α)(□α ⊃ ○)

Recalling that the Rule of Implication transforms a conditional, "□ ⊃ Δ," into "~□ V Δ" and applying the quantifier negation mechanism, it is easy to understand why this deviation occurs. There are no analogous logical equivalences for biconditionals.

10.5 Identity, Definite Descriptions, and Properties of Relations

10.5.1 Identity Statements

The addition of "=" to the vocabulary of PL increases the expressive power of PL substantially. It is easy to say explicitly that items are the same by using the special predicate "=" for identity, or say they are distinct by saying that the items are not identical. The grammar of "=" is the familiar grammar where "=" is flanked by individual terms:

$$\tau_1 = \tau_2.$$

The negation of an identity statement is written " $\tau_1 \neq \tau_2$."

Note: Using different variables does NOT express that there is more than one thing of the specified sort. Thus, "(∃x)Fx & (∃y)Fy" says only "There is at least one F and there is at least one F" (note: it does not say " ... and there exists at least one <u>other</u> F." "(∃x)Lax" translates "Ann loves someone," not "Ann loves someone <u>else</u>" (that someone Ann loves may be Ann herself). To symbolize "Ann loves someone <u>else</u>" in PL we write

   (∃x)(Lax & x ≠ a).

Numerical Statements

With the addition of identity to our symbolic language, any statement affirming that a specific finite number of As are Bs can (in theory) be expressed.

Examples

There are 0 (no) □s.	~(∃x)□x
There is at least one □.	(∃x)□x
There is at most one □.	(∀x)(∀y)[(□x & □y) ⊃ x = y]
There is exactly one □.	(∃x)[□x & (∀y)(□y ⊃ y = x)]
There is at least one □ that is a ○.	(∃x)(□x & ○x)
At most one □ is a ○.	(∀x)(∀y){[(□x & ○x) & (□y & ○y)] ⊃ x = y}
Exactly one □ is a ○.	(∃x){(□x & ○x) & (∀y)[(□y & ○y) ⊃ y = x]}
There are at least two □s.	(∃x)(∃y)[(□x & □y) & x ≠ y]
There are at most two □s.	(∀x)(∀y)(∀z){[□x & □y & □z] ⊃ (x = y ∨ x = z ∨ y = z)}
There are exactly two □s.	(∃x)(∃y){(□x & □y & x ≠ y) & (∀z)[□z ⊃ (z = x ∨ z = y)]}
There are at least two □s that are ○s.	(∃x)(∃y)[□x & □y & x ≠ y & ○x & ○y]
There are at most two □s that are ○s.	(∀x)(∀y)(∀z){[□x & ○x & □y & ○y & □z & ○z] (x = y ∨ x = z ∨ y = z)}
There are exactly two □s that are ○s.	(∃x)(∃y){□x & ○x & □y & ○y & x ≠ y & (∀z)[(□z & ○z) ⊃ (z = x ∨ z = y)]}

104

Superlatives

The meaning of superlative forms can be expressed using only the comparative form of the same word by using the identity predicate. For example, for a given UD, we can translate "FloJo is the fastest sprinter" as

$$(\forall x)[(Sx \,\&\, x \neq f) \supset Ffx],$$

where "Sx" symbolizes "x is a sprinter" and "Fxy" symbolizes "x is faster than y." In general, where "Fxy" symbolizes "x is more F than y is," we have:

a is the most F	$(\forall x)(x \neq a \supset Fax)$
a is the least F	$(\forall x)(x \neq a \supset Fxa)$

Where we have a second-place predicate "Lxy" symbolizing "x is less L than y is," we may translate:

a is the most L	$(\forall x)(x \neq a \supset Lxa)$
a is the least L	$(\forall x)(x \neq a \supset Lax)$

10.5.2 Definite Descriptions

This is the analysis of definite descriptions ("in the singular," i.e., singular definite noun phrases, which have the form "The □") due to Bertrand Russell. Russell gave a contextual definition of "the."

Russell's theory of definite descriptions takes as its starting point the observation that definite singular NPs can occur in two basic contexts:

(1) The □ is a ○

and

(2) The □ exists.

Sentences of the form (1) are understood as saying "There is exactly one □ and it is ○":

$$(\exists x)[\Box x \,\&\, (\forall y)(\Box y \supset y = x) \,\&\, \bigcirc x].$$

Sentences of the form (2) are understood as saying "There exists exactly one □":

$$(\exists x)[\Box x \,\&\, (\forall y)(\Box y \supset y = x)].$$

Other examples of English sentences that can be translated into PL as definite descriptions are possessives: "Arnie's brother," "your dog" ("the brother of Arnie," "the dog that belongs to you").

10.5.3 Properties of Binary Relations

Many place predicates are sometimes called "relation symbols" and are taken to stand for relations. Binary (second-place) relations have certain basic properties.

Assuming a particular UD:

A relation R is *reflexive* in UD iff (∀x) Rxx

A relation R is *symmetric* in UD iff (∀x)(∀y)(Rxy ⊃ Ryx)

A relation R is *transitive* in UD iff (∀x)(∀y)(∀z)[(Rxy & Ryz) ⊃ Rxz]

A relation R is *irreflexive* in UD iff (∀x) ~Rxx

A relation R is *nonreflexive* in UD iff R is neither reflexive nor irreflexive.

A relation R is *asymmetric* in UD iff (∀x)(∀y) (Rxy ⊃ ~ Ryx)

A relation R is *antisymmetric* in UD iff (∀x)(∀y)[(Rxy & Ryx) ⊃ x = y]

A relation R is *nonsymmetric* in UD iff R is neither symmetric nor asymmetric.

A relation R is *intransitive* in UD iff (∀x)(∀y)(∀z)[(Rxy & Ryz) ⊃ ~ Rxz]

A relation R is *nontransitive* in UD iff R is neither transitive nor intransitive.

A relation R is an *equivalence relation* in UD iff R is reflexive, symmetric, and transitive.

Examples

"=" is reflexive, symmetric, and transitive.

"≤" is reflexive and transitive, but not symmetric.

"<" is transitive, irreflexive, and asymmetric.

106

CHAPTER 11

Predicate Logic: Semantics

11.1 Basic Semantic Concepts: Interpretations

The semantics for SL proceeded by giving the truth conditions of all molecular sentences in terms of the truth values of their constituents. Semantics for PL is complicated by the fact that the constituents of quantified sentences are open wffs, and open wffs do not have truth values.

A semantics for PL is presented based on the *substitutional interpretation* of the quantifiers. A formal semantics for PL will also be given for *a referential* (or *objectual*) *interpretation* of the quantifiers.Semantics for SL was given in terms of **TV** assignments. For the semantics of PL, interpretations play the same role. An interpretation in part assigns denotations to all the non-logical constants of PL.

An *interpretation*, $\mathcal{I}$, consists of (1) a non-empty *domain, $\mathcal{D}$*, as the universe of discourse; (2) an assignment that (a) associates with each individual constant of PL an object in $\mathcal{D}$, (b) associates with each n-place predicate (n ≥ 1) of PL an n-ary relation among the objects of $\mathcal{D}$, and (c) assigns to each sentence letter of PL one of the truth values.

The domain $\mathcal{D}$ is a non-empty set of objects. The assignment of individual constants to objects in $\mathcal{D}$ is an assignment of an object in $\mathcal{D}$ for the constant to name. (No name may be assigned to more than one object: doing so would result in an equivocation that could lead to a fallacy. But a given object may be assigned more than one name.)

107

The assignment of n-ary relations among objects in $\mathcal{D}$ to n-place predicates means simply that subsets of objects in $\mathcal{D}$ are assigned to first-place predicates (these sets of objects in $\mathcal{D}$ are just the objects the predicate is true of — for example, the extension of the property of being a dog is the set of dogs, the extension of the property of being brown is the set of brown objects); sets of ordered pairs of objects of $\mathcal{D}$ (the extensions of binary relations) are assigned to second-place predicates—the extension of the binary relation of loving is the set of all ordered pairs of objects such that the first loves the second; etc.

We suppose that the individual constants of PL are listed in the following order: a, b, ..., v, a_1, b_1, ..., v_1, a_2, b_2, ... This assumption allows us to be quite definite when we come to select a particular constant: we will be able to speak of that the first individual constant not occurring in a given sentence.

The concept of a sentence of PL being *true under an interpretation* can be recursively defined as follows. Let $\square$, $\bigcirc$, Δ be any sentences of PL; α is any variable of PL and $\square\alpha/\beta$ is that the result of replacing all free occurrences of α in $\square$ by occurrences of β, where β is the first individual constant not occurring in $\square$. Then we can recursively define the concept of true on an interpretation $\mathcal{S}$ as follows:

<u>Truth of a sentence on an interpretation $\mathcal{S}$</u>

    (1) If $\square$ is a sentence letter, then $\square$ is true on $\mathcal{S}$ iff $\mathcal{S}$ assigns the **TV** t to $\square$.

    (2) If $\square$ is atomic and is an n-place predicate followed by n individual constants, then $\square$ is true on $\mathcal{S}$ iff the objects that $\mathcal{S}$ assigns to the individual constants of $\square$ are related (when taken in the order in which their corresponding constants occur in $\square$) by the relation that $\mathcal{S}$ assigns to the predicate of $\square$.

    (3) If $\square$ is atomic and has the form $\beta_1 = \beta_2$ (an identity sentence), then $\square$ is true on $\mathcal{S}$ iff the object $\mathcal{S}$ assigns to β_1 is the same object as that which $\mathcal{S}$ assigns to β_2.

    (4) If $\square = \sim \bigcirc$, then $\square$ is true on $\mathcal{S}$ iff $\bigcirc$ is not true under $\mathcal{S}$.

    (5) If $\square = (\bigcirc \& \Delta)$, then $\square$ is true on $\mathcal{S}$ iff $\bigcirc$ is true on $\mathcal{S}$ and Δ is true on $\mathcal{S}$.

(6) If $\Box = (\bigcirc \lor \Delta)$, then $\Box$ is true on $\mathcal{I}$ iff $\bigcirc$ is true on $\mathcal{I}$ or Δ is true on $\mathcal{I}$, or both.

(7) If $\Box = (\bigcirc \supset \Delta)$, then $\Box$ is true on $\mathcal{I}$ iff either $\bigcirc$ is not true on $\mathcal{I}$ or Δ is true on $\mathcal{I}$, or both.

(8) If $\Box = (\bigcirc \equiv \Delta)$, then $\Box$ is true on $\mathcal{I}$ iff either $\Box$ and Δ are both true on $\mathcal{I}$ or neither are true on $\mathcal{I}$.

This gives the definition of "truth under an interpretation" for all sentences except quantified sentences. Different clauses for the quantifiers give different interpretations of quantified sentences.

11.2 Substitutional Interpretation of the Quantifiers

The substitutional interpretation of the quantifiers requires a special condition on interpretations: each object in domain, $\mathcal{D}$, must be assigned at least one name.

To the definition of "true under an interpretation $\mathcal{I}$," the following clauses are added for quantified sentences.

(9) If $\Box = (\forall\alpha) \Delta$, then $\Box$ is true on $\mathcal{I}$ iff every sentence (every substitution instance) $\Delta \alpha/\beta$ is true on $\mathcal{I}$.

(10) If $\Box = (\exists\alpha) \Delta$, then $\Box$ is true on $\mathcal{I}$ iff some (at least one) sentence (substitution instance) $\Delta \alpha/\beta$ is true on $\mathcal{I}$.

$\Box$ is *false* on $\mathcal{I}$ iff $\Box$ is not true on $\mathcal{I}$.

Explanation: Given an interpretation, $\mathcal{I}$, a universally quantified sentence "$(\forall x) (... x...)$" is true on that interpretation just in case every substitution instance "$(... \beta_i ...)$" of that quantified sentence is true on $\mathcal{I}$, for every name β_i of PL. Given an interpretation, $\mathcal{I}$, an existentially quantified sentence "$(\exists x)(...x ...)$" is true on $\mathcal{I}$ just in case at least one substitution instance of that quantified sentence, "$(... \beta_i ...)$" is true on $\mathcal{I}$, for at least one name β_i of PL. Given that all members of $\mathcal{D}$ have names assigned to them, if all / some substitution instances of a quantified sentence are true, then the respective universally / existentially quantified sentence is true.

11.2.1 Truth Functional Expansions

For a domain (universe of discourse: UD) of a given finite size, where each object in that domain is assigned a name, a quantified sentence can be replaced by truth functional expansions of all its substitution instances and the result will be equivalent to the quantified sentence. This is a direct consequence of the substitutional interpretation of the quantifiers. Let the individual constants that name each of the objects in the UD be β_1, β_2, ..., β_n. Then the truth functional expansions for quantified sentences are:

A universally quantified sentence, $(\forall\alpha)\ \square$, can be replaced by a <u>conjunction</u> of substitution instances: $\square\alpha/\beta_1$ & $\square\alpha/\beta_2$ & ... & $\square\alpha/\beta_n$. That is, "$(\forall\alpha)(...\ \alpha\ ...)$" is expanded as

$(...\ \beta_1\ ...)$ & $(...\ \beta_2\ ...)$ & ... & $(...\ \beta_n\ ...)$.

In a UD of only one object, "$(\forall\alpha)(...\ \alpha\ ...)$" is expanded as

$(...\ \beta_1\ ...)$.

An existentially quantified sentence, $(\exists\alpha)\ \square$, can be replaced by a <u>disjunction</u> of substitution instances: $\square\alpha/\beta_1$ V$\square\alpha/\beta_2$ V... V$\square\alpha/\beta_n$. That is, "$(\exists\alpha)(...\ \alpha\ ...)$" is expanded as

$(...\ \beta_1\ ...)$ V $(...\ \beta_2\ ...)$ V ... V $(...\ \beta_n\ ...)$.

In a UD of only one object, "$(\exists\alpha)(...\ \alpha\ ...)$" is expanded as

$(...\ \beta_1\ ...)$.

Where there are multiple quantifiers in a wff, the quantifier that is the main logical operator is expanded first, then each sentence of the resulting expansion is in turn expanded, etc. until a molecular sentence is obtained that contains no quantifier. Alternatively, list the substitution instances of the quantified sentence to be expanded. Then form the conjunction or disjunction of these instances according to whether the quantifier that is the main logical operator is universal or existential. Then do the same, if necessary, with each constituent of the expansion that resulted.

<u>Examples</u>

Suppose a UD of one object, and $\mathcal{D}$ assigns that object to that the name "a." Then

110

Quantified sentence	Truth Functional Expansion
(∀x) Fx	Fa
(∃x) Fx	Fa
(∀x)(Fx ⊃ Gx)	Fa ⊃ Ga
(∃x)(Fx & Gx)	Fa & Ga
(∀x)(∃y)Fxy	Faa

Suppose a UD of two objects, and ℭ assigns those objects to the names "a" and "b" respectively.

Quantified sentence	Truth Functional Expansion
(∀x) Fx	Fa & Fb
(∃x) Fx	Fa VFb
(∀x)(Fx ⊃ Gx)	(Fa ⊃ Ga) & (Fb ⊃ Gb)
(∃x)(Fx & Gx)	(Fa & Ga) V(Fb & Gb)
(∀x)(∃y)Fxy	(∃y)Fay & (∃y)Fby
	(Faa VFab) & (Fba VFbb)
(∃x)[Ox & (∀y)(Ey ⊃ Gxy)	[Oa & (∀y)(Ey ⊃ Gay)] V[Ob & (∀y)(Ey ⊃ Gby)]

[Oa & {(Ea ⊃ Gaa) & (Eb ⊃ Gab)}] V[Ob & {(Ea ⊃ Gba) & (Eb ⊃ Gbb)}]

Truth functional expansions are used to construct interpretations that demonstrate certain things. For example, an interpretation on which a sentence is true shows that the sentence is not logically false in PL. Such interpretations can be constructed simply by assigning truth values to the atomic sentences of the expansions, treating distinct atomic sentences as (distinct) sentence letters.

Where the sentences of a problem consist only of <u>monadic (first-place) predicates</u>, there is a decision procedure (due to Bernays and Schönfinkel): a sentence that contains k distinct monadic predicates is logically true in PL iff the sentence is true on every interpretation with a UD containing 2^k members. So if we have a sentence consisting of only k monadic predicates, we can expand the sentence for a set of 2^k selected individual constants, and determine by truth table methods (truth trees, short-cut truth tables) whether the sentence is

111

true on every interpretation with a UD that is the same size as the number of selected constants. If the sentence is true on every interpretation with a UD that consists of 2^k objects, then by the Bernays-Schönfinkel result, the sentence is logically true in PL.

11.2.2 Limitations Of The Substitutional Interpretation

For the substitutional interpretation of the quantifiers and the method of truth functional expansions to be satisfactory, every object in the UD must have an individual constant assigned to it by the interpretation. Otherwise, some unnamed object in the UD might fail to be in the extension assigned by an interpretation to the predicate "Fx"; yet "(∀x)Fx" might be true on the substitutional interpretation because all its substitution instances (all the objects in UD that have names) do have the property. It is desirable to have an interpretation of the quantifiers that allows us to apply our logic even to cases where objects may fail to have names. In the case of the real numbers, for example, a proof by Cantor shows that we cannot have enough individual constants in PL to put them into 1-1 correspondence with the real numbers, so some real numbers must go unnamed.

11.3 Referential (Objectual) Interpretation of the Quantifiers

In this interpretation of the quantifiers, there is no restriction that all objects in the UD must have names assigned to them.

We introduce the concept of a "β-variant of an interpretation." Let $\mathcal{I}_1$ and $\mathcal{I}_2$ be interpretations of PL. Let β be an individual constant of PL. Then:

$\mathcal{I}_2$ is a β-variant of $\mathcal{I}_1$ iff either

    (a) $\mathcal{I}_2$ is exactly the same as $\mathcal{I}_1$; or

    (b) $\mathcal{I}_2$ differs from $\mathcal{I}_1$ <u>at most</u> in respect to what object they assign to β.

    Note that for one interpretation to be a β-variant of another, they must have exactly the same UD; they must assign the same objects to the same individual constants (with the possible exception that $\mathcal{D}_2$ may assign a different object of $\mathcal{D}$ to β than $\mathcal{D}_1$ assigns to β); they must assign the same sets of objects to the predicates and the same **TV**s to the sentence letters.

To the definition of "true under an interpretation" given above in section 11.1, we add clauses for the quantifiers:

(9) If $\square$ = (∀α) Δ, then $\square$ is true on $\mathcal{D}$ iff Δα/β is true on every β-variant of $\mathcal{D}$.

(10) If $\square$ = (∃α) Δ, then $\square$ is true on $\mathcal{D}$ iff Δα/β is true on at least one β-variant of $\mathcal{D}$.

$\square$ is *false* on $\mathcal{D}$ iff $\square$ is not true on $\mathcal{D}$.

Explanation:

The assignment of objects in UD to a constant cannot vary within a given interpretation, because the interpretation $\mathcal{D}$ has fixed such an assignment. Instead, different interpretations that vary in the desired way are considered. The object that $\mathcal{D}$ assigns to the constant is considered, and then we consider what objects every other interpretation just $\mathcal{D}$ in every other respect assigns to the name. By considering a sequence of β-variant interpretations, the constant will name at some place in the sequence each and every object in the UD. In this way, the **TV** of the substitution instance is evaluated where each and every object in UD is taken as the denotation of the name. If the sentence Δα/β is true on all of these β-variant interpretations of $\mathcal{D}$, (∀α)Δ is true on $\mathcal{D}$, since the condition expressed in Δ will be true of every object in the UD. If the sentence Δα/β is true on at least one β-variant of $\mathcal{D}$, then (∃α)Δ is true on $\mathcal{D}$, since the condition expressed by Δ will be true of at least one object in the UD.

11.4 Referential Interpretations Given in Terms of Satisfaction

Another common way to address the problem of giving a definition of "true under an interpretation" in light of the fact that the constituents of quantified sentences are not themselves sentences, is to define "satisfaction on an interpretation" for all wffs (open and closed), and then to define "true on an interpretation" for sentences in terms of "satisfaction." This will be done in terms of the notion of the satisfaction of a wff by a denumerable sequence of objects.

A *sequence* of n objects is an ordered n-tuple of objects. A sequence Σ is the same sequence as the sequence Σ' iff Σ and Σ' have exactly the same number of elements and the first element of Σ is the same as the first element of Σ', the second element of Σ is the same as the second element of Σ', etc. So $<3, 2, 3> \neq <3, 3, 2>$; and $<2, 3> \neq <2, 2, 3>$.

Interpretations are the same as before. Let $\mathcal{I}$ be any interpretation with $\mathcal{D}$ as its universe of discourse, $\square$ any sentence of PL, Σ any denumerable sequence of objects that are elements of $\mathcal{D}$. The variables are understood as being ordered as before.

The satisfaction of a wff $\square$ by a sequence Σ on a given interpretation $\mathcal{I}$

(1) If $\square$ is a sentence letter of PL, then Σ satisfies $\square$ on I iff I assigns $\square$ that the **TV** t.

(2) If $\square$ is an atomic sentence but not a sentence letter, then $\square$ is of the form $\Delta \beta_1, \beta_2, ..., \beta_n$, where Δ is an n-place predicate and $\beta_1, \beta_2, ..., \beta_n$ are (not necessarily distinct) individual constants. $\mathcal{I}$ assigns to each constant an element of $\mathcal{D}$: let the members of $\mathcal{D}$ assigned to $\beta_1, \beta_2, ..., \beta_n$ be $o_1, o_2, ..., o_n$, respectively. Then Σ satisfies $\square$ iff the ordered n-tuple $<o_1, o_2, ..., o_n>$ is a member of the set of ordered n-tuples that $\mathcal{I}$ assigned to the predicate Δ.

(3) If $\square$ is an atomic wff of the form $\Delta \alpha_1, \alpha_2, ..., \alpha_n$, where Δ is an n-place predicate and $\alpha_1, \alpha_2, ..., \alpha_n$ are (not necessarily distinct) variables, then to each variable α_i an object in Σ is assigned as follows: if α_i is the kth variable in our enumeration of the variables of PL, then α_i is assigned the kth member of Σ. Let the objects in Σ assigned to the variables $\alpha_1, \alpha_2, ..., \alpha_n$ be $o_1, o_2, ..., o_n$ (not necessarily distinct). Then Σ satisfies $\square$ iff the ordered n-tuple $<o_1, o_2, ..., o_n>$ is a member of the set of ordered n-tuples $\mathcal{I}$ assigned to the predicate Δ.

(4) If is an atomic wff of the form $\Delta \tau_1, \tau_2, ..., \tau_n$, where $\tau_1, \tau_2, ..., \tau_n$ are terms (i.e., each τ_i is either an individual constant or a variable) and is Δ an n-place predicate, then

114

if τ_1 is a constant, it is assigned a member of $\mathcal{D}$ (call that the object o_i) by $\mathcal{D}$; if τ_i is a variable, it occurs in our enumeration of the variables of PL; suppose it is the kth variable in the enumeration. Then τ_i is assigned the kth member of the sequence Σ: call this object o_i. So if τ_i is a constant, o_i is the object from $\mathcal{D}$ assigned to it by $\mathcal{D}$. And if τ_i is a variable, then o_i is the object of $\mathcal{D}$ assigned to it by looking at that the enumeration of the variables to find which variable in the enumeration τ_i is — say it's the nineth; and then examining Σ and assigning the nineth member of Σ to τ_i (in this case, o_i is an object in the $\mathcal{D}$ since all members of are members of $\mathcal{D}$). Then Σ satisfies $\square$ iff $<o_1, o_2, ..., o_n>$ is a member of the set of ordered n-tuples assigned to the predicate Δ.

(5) If $\square$ = ~Δ, then Σ satisfies $\square$ iff Σ does not satisfy Δ.

(6) If $\square$ = $(\Delta \ \& \ \bigcirc)$, then Σ satisfies $\square$ iff Σ satisfies Δ and Σ satisfies $\bigcirc$.

(7) If $\square$ = $(\Delta \ \lor \bigcirc)$, then Σ satisfies $\square$ iff Σ satisfies Δ or Σ satisfies $\bigcirc$ (or both).

(8) If $\square$ = $(\Delta \supset \bigcirc)$, then Σ satisfies $\square$ iff Σ does not satisfy Δ or Σ satisfies $\bigcirc$ (or both).

(9) If $\square$ = $(\Delta \equiv \bigcirc)$, then Σ satisfies $\square$ iff Σ satisfies both Δ and $\bigcirc$ or Σ satisfies neither Δ nor $\bigcirc$.

(10) If $\square$ = $(\forall\alpha_k)\Delta$, where α_k is the kth variable in the enumeration, then Σ satisfies $\square$ iff every sequence of members of $\mathcal{D}$ that differs from Σ at most in the kth member satisfies Δ. I.e., Σ satisfies $\square$ iff no matter which object from $\mathcal{D}$ is assigned to the kth member in Σ, the resulting sequence (which differs from Σ at most at its kth place) satisfies Δ.

(11) If $\square$ = $(\exists\alpha_k)\Delta$, where α_k is the kth variable in the enumeration, then Σ satisfies $\square$ iff at least one sequence of members of $\mathcal{D}$ that differs from Σ at most in its kth place satisfies Δ.

A sentence $\square$ of PL is *true* on a given interpretation $\mathcal{D}$ iff <u>every</u> denumerable sequence of members of $\mathcal{D}$ satisfies $\square$.

A sentence ☐ of PL is *false* on a given ☉ iff <u>no</u> denumerable sequence of members of 𝒟 satisfies ☐.

11.5 Semantic Properties of Expressions of PL

Semantic properties of expressions of PL are defined in ways analogous to those for SL, with interpretations playing the role of truth value assignments.

To determine whether certain sentences or sets of sentences of PL have certain semantic properties, one might attempt to construct an interpretation that will do the job required. (What is needed from an interpretation will be indicated for each semantic property.) One may attempt to construct an appropriate interpretation either by specifying a UD and assigning extensions for all the nonlogical vocabulary (individual constants, predicates) occurring in the sentence(s) or by using truth functional expansions for selected UDs of various sizes (usually consisting of one, two, or three objects); or one might construct an interpretation where all nonlogical vocabulary is assigned a meaning by giving a symbolization key for translating them into English.

11.5.1 Consistency

A sentence of PL ☐ is *consistent* iff there is at least one interpretation on which it is true.

A set of sentences of PL{☐$_1$, ☐$_2$, ..., ☐$_n$} is *consistent* if there is at least one interpretation on which each member of the set is true.

A set of PL sentences is *inconsistent* iff it is not consistent (i.e., iff there is no interpretation on which all members of the set are true).

To show that a sentence (or a set of sentences) of PL is consistent, construct an interpretation on which the sentence (or all members of the set of sentences) is true.

To show that a sentence (set of sentences) is inconsistent, prove that there is no interpretation on which the sentence (all members of the set of sentences) is true.

116

11.5.2 Logical Truth

A sentence of PL □ is *logically true* in PL iff □ is true under every interpretation.

To establish that a sentence is logically true in PL, prove that there is no interpretation on which it is false. To establish that a sentence is <u>not</u> logically true in PL, construct an interpretation on which the sentence is false.

11.5.3 Logical Falsity

A sentence □ of PL is *logically false* in PL iff □ is false on every interpretation.

To show that a sentence is logically false, prove that there is no interpretation on which it is true. To show that a sentence is <u>not</u> logically false, construct an interpretation on which it is true.

11.5.4 Contingency

A sentence □ of PL is *contingent* in PL iff □ is neither logically true nor logically false in PL.

11.5.5 Logical Equivalence

Sentences □, Δ are *logically equivalent* in PL iff there is no interpretation on which □ and Δ have different truth values.

To show that two sentences are logically equivalent, prove that there is no interpretation on which their truth values differ. To show that two sentences are <u>not</u> logically equivalent, construct an interpretation on which they have different truth values.

11.5.6 Logical Consequence

A sentence, Δ, is a *logical consequence* of a set of sentences Γ iff there is no interpretation under which all members of Γ are true and Δ is false.

To show that a sentence, Δ is a logical consequence of a set of sentences, Γ, prove that there is no interpretation on which all members of Γ are true and Δ is false. To show that a sentence, Δ, is <u>not</u> a logical consequence of a set of sentences, Γ, construct an interpretation on which all members of Γ are true but Δ is false.

11.5.7 Validity

An argument

$$\Box_1, \Box_2, ..., \Box_n \, / \, \therefore \Delta$$

is *valid* iff there is no interpretation on which all its premises, $\Box_1, \Box_2, ..., \Box_n$, are true and its conclusion, Δ, is false.

To show that an argument is valid, prove that there is no interpretation on which all its premises are true and its conclusion false. To show that an argument is <u>not</u> valid, construct an interpretation on which all its premises are true and its conclusion is false.

CHAPTER 12

Predicate Logic: Truth Trees

12.1 Truth Tree Rules for Quantified Sentences

The tree rules and the tree structure for sentence logic are carried over to predicate logic. Additional rules are introduced for quantifiers and their negations and for identities.

12.1.1 Quantifier Rules

Universal Quantifier Decomposition: Rule

$(\forall\alpha)\ \square$

$\square\ \alpha/\beta$

If a universally quantified sentence, $(\forall\alpha)$ (... α ...) occurs as a full sentence at a node of a tree, perform the following procedure: For any or all names that occur in any sentence on that branch, write the substitution instance (... β ...) beneath every open branch through $(\forall\alpha)$ (... α ...), unless (... β ...) already occurs in the branch. If no name occurs on the branch, pick any name and write a substitution instance using it beneath every open branch through the quantified sentence. Do not check the sentence $(\forall\alpha)$ (... α ...). Optionally: in addition, list the names occurring in the substitution instances to the right of the quantified sentence to which the decomposition rule has been applied.

119

Existential Quantifier Decomposition: ∃ Rule

(∃α) □ ✔

□ (α/β)

where β is an individual constant new to the branch or branches (that is, β has not occurred in any sentence on any branches through the quantified sentence).

If an existentially quantified sentence, (∃α) (... α ...) occurs as the entire sentence at a node of an open branch of a tree, perform the following procedure: for each open branch through the quantified sentence, pick a name, β, <u>new to the branch</u> and write the substitution instance (... β ...) at the bottom of the branch. Put a check mark to the right of the quantified sentence.

12.1.2 Negated Quantifier Rules

Negated Universal Quantifier Rule: ~ ∀ Rule

~(∀α) □ ✔

(∃α) ~ □

Negated Existential Quantifier Rule: ~ ∃ Rule

~ (∃α) □ ✔

(∀α) ~□

Both negation rules work like the decomposition tree rules of sentence logic: write the result of applying the rule at the bottom of every open branch through the sentence to which the rule is applied.

12.2 Identity Rules

Identity Decomposition Rule: = D

Let "□ β₂ // β₁" represent the result of replacing one or more occurrences of the individual constant β₁ in □ by occurrences of the constant β₂. Then the =D rule is

β₁ = β₂	(1) β₁ = β₂	(1) β₁ = β₂
□	(2) ... β₁ ...	(2) ... β₂ ...
□ β₁ // β₂	... β₂ ...	... β₁ ...

120

Note: <u>neither sentence is checked after the application of the rule</u>. This rule says that if an open branch of a tree has a sentence on a full line of the form (1) "$\beta_1 = \beta_2$," and a line (2) in which one of those names occurs, write at the bottom of every open branch through those lines a like (2) except that one or more occurrences of one name has been replaced by occurrences of the other name.

Negated Identity Rule: ~ =

Close any branch of a tree containing $\beta \neq \beta$.

12.3 Strategies for Constructing Truth Trees with Quantified Sentences

Wherever possible:

- Apply rules to sentences whose decomposition does not require branching.

- Apply rules to sentences whose decomposition results in one or more branches closing.

- Apply rules to existentially quantified sentences before applying rules to universally quantified sentences.

12.4 Definitions of Tree Properties

Because universally quantified sentences are not checked, it is possible for trees to have open branches that never become complete. In some cases, a branch may become infinitely long. The definition of a completed open branch is revised to account for this possibility.

A branch of a truth tree for PL is a *completed open branch* iff it is a finite open branch (an open branch with a finite number of nodes (full lines)), and each sentence occurring on that branch is either (i) a literal; or (ii) a sentence other than a universally quantified sentence that has been decomposed (checked); or (iii) a universally quantified sentence $(\forall \alpha) \square$, and where $\square$ α/β occurs on that branch for each constant β occurring in a sentence on that branch, and where at least one such instance $\square$ α/β does occur on the branch; or (iv) a sentence of the form $\beta_1 = \beta_2$ and where the branch also contains, for every literal $\square$ containing β_1 on that branch, every sentence $\square$ $\beta_2 /\!/ \beta_1$

which can be obtained from □ by the =Rule and which is not of the form β = β.

A *closed branch* is a branch on which some sentence and its negation both occur.

An *open branch* is a branch which is not closed.

A *completed tree* is a tree on which each branch is either closed or is a completed open branch.

A *closed tree* is a tree on which every branch is closed.

12.5 Semantic Properties as Characterized by Truth Trees

These definitions are analogous to those for sentence logic. To test for a given property, the same set of initial sentences is selected for the tree; the same outcome, namely a closed tree, indicates that no interpretation can be constructed making all the initial sentences true.

A sentence □ of PL is *logically true* in PL iff the set {~ □} has a closed truth tree.

A sentence □ of PL is *logically false* in PL iff the set {□} has a closed truth tree.

A sentence □ of PL is *contingent* (*logically indeterminate*) in PL iff it is neither logically true in PL nor logically false in PL.

Sentences □, Δ of PL are *logically equivalent* in PL iff {~(□ ≡ Δ)]} has a closed truth tree.

A finite set Γ of sentences of PL *entails* a sentence Δ in PL iff Γ ∪ {~Δ} has a closed truth tree.

An argument

is *valid* in PL iff {□₁, □₂, ..., □ₙ, ~Δ) has a closed truth tree.

122

CHAPTER 13

Predicate Logic: Derivations

13.1 Inference Rules of Predicate Logic: Introduction and Elimination Rules

The same derivation system for sentence logic is carried over to predicate logic, with quantifier rules for a "Fitch-style" system added. The resulting derivation system is called "PD." Different texts adopt different rules of inference for quantifiers. Several alternative rules for quantifiers are presented in 13.2.

13.1.1 Quantifier Rules

Universal Elimination: ∀ Elim (Universal Instantiation: UI)

$$\text{j.} \quad \Big| \quad (\forall \alpha)\,\square$$

$$\cdots$$

$$\triangleright \quad \text{k.} \quad \Big| \quad \square\ \alpha/\beta\ \text{j, }\forall\text{Elim}$$

This rule permits us to infer from "Everything is so and so" that any instance is so and so. To apply the rule, delete the universal quantifier that is the main logical operator of the sentence, replace every occurrence of the variable linked to it by occurrences of the individual constant selected to be the instantiating constant.

Universal Introduction: ∀Int (Universal Generalization: UG)

For universal introduction, the form is simple: from a substitution instance of it, a universally quantified statement may be inferred (provided two conditions are satisfied).

j. | $\Box$ α/β

...

▷ k. | $(\forall\alpha)\Box$ j, $\forall$Int

provided:

(i) $\Box$ does not occur in any undischarged assumption (including premises);

(ii) $\Box$ does not occur in $(\forall\alpha)$ $\Box$.

The intuitive idea of universal introduction is that if some property (or condition) can be established about any object, any arbitrarily selected object, then we may conclude that that property (or condition) holds of every object. We may universally generalize from an instance only if the information we have about the object mentioned in the instance is information that we have about any other object that could be cited. Thus, if the name in the instance were replaced by any other name, the (semantic) validity of the inference would not be affected. This is the consideration that motivates the restrictions on the rule.

Note: condition (ii) requires that all occurrences of the constant being generalized on be replaced throughout by occurrences of a variable that does not already occur in the instance, which are then bound by an initial universal quantifier whose scope is the entire expression. Thus the following is an <u>erroneous</u> application of the $\forall$Int Rule:

1. | $(\forall x)\,Lxx$ | Premise | Every number is less than or equal to itself

2. | Laa | 1, $\forall$Elim | One is less than or equal to one

3. | $(\forall x)\,Lxa$ | 2, $\forall$Int ERROR! | Every number is less than or equal to one.

The rule of $\forall$Int is easy to apply: select an *instantiating constant*, β, occurring in $\Box$; verify that β doesn't occur in any premise or assumption that has not been discharged; replace <u>all</u> occurrences of β by any variable, α, that does not occur in $\Box$; prefix a universal quantifier "$(\forall\alpha)$" whose scope is the whole expression.

124

Existential Introduction: ∃Int (Existential Generalization: EG)

This rule permits us to infer an existentially quantified sentence from an instance of it. For example, in natural language we infer that someone is wealthy from Alice is wealthy. Thus, in our system of logic, the form of the rule is:

j.   □ (α/β)

       ...

k.   (∃α) □   j, ∃Int

Note: Int permits the replacement of <u>one or more</u> occurrences of a name by a variable not occurring in □ (α/β) and introducing an existential quantifier whose scope is the entire expression. To apply this rule to a sentence □ containing one or more occurrences of a name, β: replace one or more occurrences of β by any individual variable, α, which does not occur in □, and prefix "(∃α)" making this quantifier the main logical operator of the resulting sentence.

Existential Quantifier Elimination: ∃Elim (Existential Specification: ES)

The rule of existential quantifier elimination permits us to <u>assume</u> an instance if given an existentially quantified sentence on a line of a derivation. If from this assumption (and perhaps other information available in earlier lines) we can infer some sentence, then (if three conditions are satisfied) we may then infer the sentence from the existentially quantified sentence (together with that other information).

The form of the rule is this:

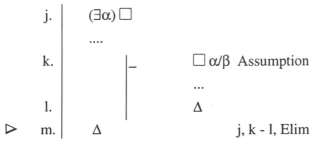

125

provided:

(i) β does not occur in any premise or undischarged assumption (except (k));

(ii) β does not occur in (∃α) □;

(iii) β does not occur in Δ.

A sentence of the form "(∃x)Fx" tells us that something has the property F but not which thing has it. So we <u>assume</u> that the object that has F is, say, *a*. Since that is all that is assumed about *a*, the object *a* can represent any object at all that has the property F. Thus instead of assuming "Fa," any other object could have been assumed to have the property F without affecting the subderivation that follows the assumption. This guarantees that whatever we establish on the assumption "Fa" we could establish from any other assumption "Fb," "Fc," etc. and ensures that we make use of no information about the object *a* except that it might have the property F. This is the rationale for the first two restrictions: they specify what is required of a satisfactory representative.

A similar rationale applies for condition (iii). For ∃Elim to be correct, a correct derivation of Δ must be possible no matter what particular object we take to be the representative any instance of (∃x) □. We know that something has the property "□." We do not know that the object *a* has the property □; "a" is merely "representing" whatever object has □.

Suppose that some result could be derived that says that *a* has some other property, say Δ; so we are able to derive "Δa":

i.		(∃x) □x	
j.	−	□ a	Assumption
k.		Δ a	

Restriction (iii) prevents us from inferring "Δ a" by ∃ Elim from i, j-k which would be an invalid inference, as the following shows:

1.	−	(∃x) Ex	Premise		Some number is even.
2.		− Ef	Assumption		Five is even.
3.		Ef	2, Reit		Five is even.
4.		Ef	1, 2 - 3, ∃ Elim	ERROR!	Five is even.

126

To apply the ∃ Elim rule correctly, an existentially quantified sentence must occur as a full line of the derivation. Then assume an instance, selecting as the instantiating constant a constant that has not occurred anywhere earlier in the derivation. This assures that restrictions (i) and (ii) are satisfied. Before closing the subderivation and justifying the last line, Δ, be sure that the instantiating constant does not occur in Δ. That is, the subderivation cannot be closed and its last line copied outside the subderivation if the constant of instantiation occurs in that line.

13.1.2 Derived Rules for Quantifiers

The following rules can be derived in PD: anything that can be derived in PD using these rules can be derived in PD without using them. They make some derivations shorter and easier, and are convenient for this reason.

Quantifier Negation: QN

$\sim(\forall\alpha)\,\square$ :: $(\exists\alpha)\sim\square$

$\sim(\exists\alpha)\,\square$ :: $(\forall\alpha)\sim\square$

Note that these are replacement rules, so can apply to subwffs within a sentence on a line of a derivation as well as to whole sentences. The mechanism for applying the QN rule is quite easy: A negation may be moved through a quantifier (moving the negation either from left of the quantifier to the right of it or from the right of the quantifier to the left of it), but when so moved, the quantifier changes to its dual (universal becomes existential; existential becomes universal).

13.1.3 Identity Rules

Identity Introduction: =Int (Reflexivity of Identity: Refl =)

▷   j. $\;\Big|\;$ $(\forall\alpha)\,\alpha = \alpha$ =Int

This rationale of this rule is that everything is identical to itself. Hence we can introduce that statement at any line of a derivation.

127

Identity Elimination: =Elim (Indiscernibility of Identicals: Ind Id)

When two names, n_1 and n_2, name the same object (so "$n_1 = n_2$" is true), then if the object named by n_1, say, has some property, it may be inferred that the object named by n_2 also has that property, since it is one and the same object as that named by n_1. This is the rationale for the =Elim rule, the form of which is:

$$
\begin{array}{ll}
\text{j.} & \beta_1 = \beta_2 \\
& \cdots \\
\text{k.} & \square \\
\rhd \quad \text{l.} & \square\, \beta_1 \mathbin{/\!/} \beta_2
\end{array}
\qquad
\begin{array}{ll}
\text{j.} & \beta_1 = \beta_2 \\
& \cdots \\
\text{k.} & \square \\
\rhd \quad \text{l.} & \square\, \beta_1 \mathbin{/\!/} \beta_2
\end{array}
$$

where "$\square\ \beta_1 \mathbin{/\!/} \beta_2$" means that one or more occurrences of β_1 in $\square$ have been replaced by occurrences of β_2.

13.2 Other Versions of Quantifier and Identity Rules

Quantifier rules vary from text to text. Some texts allow wffs that are not sentences to occur in lines of derivations. (In this case, the question of the truth-preserving character of the rules becomes complicated. In these cases, a completed derivation must have a sentence as its last line, and the rules will be truth-preserving in the sense that a false conclusion (last line) cannot be derived from all true premises if the rules are strictly observed.) Commonly, four quantifier rules are presented: an elimination rule for each quantifier and an introduction rule for each quantifier. The rules for Universal Elimination (Universal Instantiation) and Existential Introduction (Existential Generalization) may vary only slightly from those given above, modified so as to interact appropriately with the other two rules. By an "existential name" is understood an individual constant introduced into a derivation by an application of the rule of Existential Instantiation.

Universal Instantiation: UI

j.	$(\forall\alpha)\,\Box\alpha$		j.	$(\forall\alpha)\,\Box\alpha$	
	...			...	
k.	$\Box\,\alpha_1$	j, UI	k.	$\Box\beta$	j, UI

Universal Generalization: UG

j.	$\Box\alpha_1$		j.	$\Box\beta$	
	...			...	
k.	$(\forall\alpha)\,\Box\alpha$	j, UG	k.	$(\forall\alpha)\,\Box\alpha$	j, UG ERROR!

with the restrictions that

(i) UG cannot be used in a subderivation if the variable of instantiation α_1 occurs free in the assumption that initiates the subderivation;

(ii) UG cannot be used if $\Box\alpha_1$ contains an existential name and α_1 is free in the line where that name is introduced.

Existential Instantiation: EI

j.	$(\exists\alpha)\,\Box\alpha$		j.	$(\exists\alpha)\,\Box\alpha$	
	...			...	
k.	$\Box\beta$	j, EI	k.	$\Box\alpha$	j, EI ERROR!

with the restriction that the existential name, β, must be a name that has not occurred in any previous line.

Existential Generalization: EG

j.	$\Box\beta$		j.	$\Box\alpha_1$	
	...			...	
k.	$(\exists\alpha)\,\Box\alpha$	j, EG	k.	$(\exists\alpha)\,\Box\alpha$	j, EG

Identity Introduction: =In

This rule permits the introduction on any line of a derivation of an identity sentence using any name:

j. | $\beta = \beta$ =Int

13.3 Basic Concepts of Derivation Systems for Predicate Logic

The basic concepts of PD are analogous to those of SD.

A *derivation* in PD of a sentence of PL Δ from a set of sentences of PL, Γ, is a consecutively numbered finite, non-empty sequence of sentences of PL in which the last sentence in the sequence is Δ, and each sentence is either (i) a member of Γ, or (ii) an assumption with its scope indicated by a vertical line extending until the assumption is discharged, or (iii) an axiom of PD, or (iv) is the result of applying a rule of inference of PD to one or more of the sentences preceding it in the sequence.

A sentence, Δ, is *derivable* in PD from a set of sentences, Γ, ($\Gamma \vdash_{PD} \Delta$) iff there is a derivation in PD of Δ from Γ.

A sentence, Δ, is a *theorem* ($\vdash_{PD} \Delta$) of PD iff Δ is derivable from the empty set of premises.

Sentences $\square$, Δ are *equivalent* in PD iff $\{\square\} \vdash_{PD} \Delta$ and $\{\Delta\} \vdash_{PD} \square$ (that is, two sentences are equivalent in PD iff they are derivable from each other).

An argument of SL,

$$\square_1, \square_2, ..., \square_n / \therefore \Delta$$

is *valid* in PD iff $\{\square_1, \square_2, ..., \square_n\} \vdash_{PD} \Delta$ (that is, iff its conclusion is derivable in PD from its premises).

A set of sentences, Γ, is *inconsistent* in PD iff there is some sentence, Δ, of PL such that both $\Gamma \vdash_{PD} \Delta$ and $\Gamma \vdash_{PD} \sim\Delta$ (that is, a set of sentences Γ is inconsistent in PD iff some sentence and its negation are both derivable in PD from Γ).

13.4 Theorems of PD

Selected theorems of PD follow. The number of theorems of PD is (denumerably) infinite.

1. $(\forall x)(\forall y)\, Lxy \equiv (\forall y)(\forall x)\, Lxy$

2. $(\exists x)(\exists y)\, Lxy \equiv (\exists y)(\exists x)\, Lxy$

3. $(\exists x)(\forall y)\, Lxy \supset (\forall y)(\exists x)\, Lxy$

4. $(\forall x)(Ax \mathbin{\&} Bx) \equiv [(\forall x)\, Ax \mathbin{\&} (\forall x)\, Bx]$

5. $(\forall x)(Ax \supset Bx) \supset [(\forall x)Ax \supset (\forall x)\, Bx]$

6. $[(\forall x)\, Ax \lor (\forall x)\, Bx] \supset (\forall x)(Ax \lor Bx)$

7. $(\exists x)(Ax \mathbin{\&} Bx) \supset [(\exists x)Ax \mathbin{\&} (\exists x)Bx]$

8. $(\exists x)(Ax \lor Bx) \supset [(\exists x)Ax \lor (\exists x)\, Bx]$

9. $(\forall x)(\exists y)(Ax \mathbin{\&} By) \equiv [(\forall x)\, Ax \mathbin{\&} (\exists y)\, By]$

10. $(\forall x)(\exists y)(Ax \mathbin{\&} By) \equiv (\exists y)(\forall x)(Ax \mathbin{\&} Gy)$

11. $(\forall x)(\exists y)(Ax \lor By) \equiv [(\forall x)Ax \lor (\exists y)By]$

12. $(\forall x)(\exists y)(Ax \lor By) \equiv (\exists y)(\forall x)(Ax \lor By)$

13. $(\forall x)(\forall y)(Ax \supset Ay) \equiv [\sim (\exists x)\, Ax \lor (\forall x)\, Ax]$

14. $[(\forall x)\, Ax \supset (\forall x)Bx] \equiv (\exists x)(\forall y)(Ax \supset By)$

15. $\sim (\exists y)(\forall x)(\, Axy \equiv\, \sim Axx)$

131

CHAPTER 14

Inductive Logic

14.1 Argument by Analogy

An argument by analogy is an argument based on a comparison or analogy as follows:

(1) A comparison is made between two or more objects with respect to their sharing various properties in common.

(2) All but one of these objects is claimed to have an additional property.

(3) The conclusion is drawn that the other object has the property.

Call the object mentioned in the conclusion the "target entity" and the property mentioned in the conclusion the "target property." Call the other objects mentioned in the premises the "base entities" and the properties mentioned in the premises, except the target property, the "base properties."

The form of a simple argument by analogy can be represented as follows:

Aa & Ba & Ca & ... & Ma	Here, the base properties are A, B, C, ..., M;
Ab & Bb & Cb & ... & Mb	the base entity is a; the target property is N; and
Na / Nb	the target entity is b.

A more complex pattern of argument by analogy simply increases the number of base entities.

The *positive analogy* is the properties the base entities and the target entity share in common: A—M in the form above. The *negative analogy* is the properties not shared between the target entity and the base entities. The *degree of analogy* between entities is greater with larger positive and smaller negative analogy, and smaller with smaller positive and larger negative analogy. The *range* of the analogy is the set of entities involved in the argument by analogy. The *relevance* of one or more properties to another property is the regular or lawful connection between them. For example, horsepower of a car engine is relevant to the car's gas mileage, whereas color of the body is not.

Several rules are common for evaluating the strength of an argument by analogy.

<u>Rule of Relevance.</u> The more/less relevant the positive analogy is to the conclusion, the stronger/weaker the argument; and/or the more/less relevant the negative analogy is to the conclusion, the weaker/stronger the argument.

<u>Rule of Degree</u> The greater the degree of analogy, the stronger the argument (other considerations being equal).

<u>Rule of Range</u> The greater the range of analogy, and the more diverse its members, the stronger the argument.

<u>Rule of Strength of the Conclusion</u> The stronger/weaker the conclusion, the weaker/stronger the argument. the more/less specific the conclusion, the weaker/stronger the argument.

14.2 Induction by Simple Enumeration

Induction by simple enumeration is closely related to argument by analogy. This form of inference underlies such common examples as the following: All emeralds are green, since all emeralds observed so far have been green.

The form of an argument by induction by simple enumeration is:

Instance 1	Aa & Ba
Instance 2	Ab & Bb
...	...
Instance n	Am & Bm

All As are Bs

14.3 Mill's Methods

Mill's Methods are methods for determining the probable presence of causally necessary or sufficient conditions of a given property from observation of instances of events.

The *conditioned property* is the property whose necessary or sufficient conditions are being sought. *Possible conditioning properties* (PCPs) are those properties under consideration as being necessary or sufficient for a given conditioned property. Mill's Methods are ways of selecting necessary or sufficient causal conditions from among the PCPs on the basis of observation of various instances. The methods involve practical applications of the definitions of necessary condition and of sufficient condition.

A property P is a *necessary condition* for a property Q iff whenever Q is present, P is present; Q cannot occur without P's having occurred: if no P, then no Q.

A property P is a *sufficient condition* for a property Q iff whenever P is present, Q is present; Q can't fail to occur when P occurs: if P, then Q.

In applying Mill's Methods, it is assumed that the PCPs include the causally necessary or sufficient conditions for the conditioned property under investigation. It must also be recognized that the conclusions drawn may be overturned by examination of additional instances; these inferences are inductive.

Assume for purposes of illustration a set {A, B, C, D} of PCPs and a conditioned property E.

14.3.1 The Direct Method of Agreement

The Direct Method of Agreement is used to identify a necessary condition for a conditioned property E from among the PCPs. Any

134

property that is absent when the conditioned property E is present cannot be a necessary condition of E. If, where E is present in all observed instances, only C is present in all observed instances, then the other PCPs cannot be necessary conditions for E. Hence, if a necessary condition for E is among the PCPs under consideration, then it is C.

Elimination Principle: A property that is absent when E is present can't be a necessary condition for E.

Factor Tabulation			
Instance	Possible Conditioning Properties		Conditioned Property
	A	B	C	E
1	Present	Present	Present	Present
2	Present	Absent	Present	Present
3	Absent	Present	Present	Present
If one of the PCPs is a necessary condition for E, C is that necessary condition			

14.3.2 The Inverse Method of Agreement

The Inverse Method of Agreement is used to identify a sufficient condition for a conditioned property E from among the PCPs. Any property that is present when the conditioned property E is absent cannot be a sufficient condition of E. If, where E is absent in all observed instances, only C is absent in all observed instances, then the other PCPs cannot be sufficient conditions for E. Hence, if a sufficient condition for E is among the PCPs under consideration, then C is that sufficient condition.

Elimination Principle: A property that is present when E is absent can't be a sufficient condition for E.

135

Factor Tabulation			
Instance	Possible Conditioning Properties		Conditioned Property
	A	B	C	E
1	Present	Absent	Absent	Absent
2	Absent	Present	Absent	Absent
3	Present	Absent	Absent	Absent
If one of the PCPs is a sufficient condition for E, C is that sufficient condition			

14.3.3 The Method of Difference

The Method of Difference is used to find a sufficient condition for a conditioned property in a particular instance where it is present. That is, the Method of Difference is used to determine from among the conditions present when a particular instance of E occurs, which property is a sufficient condition for E on this occurrence. For example, being blown apart by an explosive is sufficient for death, but when an intact corpse is discovered, that sufficient condition for death will not be the one sufficient for death in this instance, for the person was not blown up; something else was sufficient for death in this case. When a given conditioned property is present, some of its sufficient conditions may be absent. Any sufficient condition present in the instance of interest is sufficient for E in that instance. The Method of Difference is used to determine which of the PCPs actually present on a particular occurrence of E is a sufficient condition for E.

Principle of Elimination: A property that is present when E is absent cannot be a sufficient condition for E.

Let "Occurrence*" be the particular occurrence of E for which a sufficient condition is sought.

136

Factor Tabulation				
Instance	Possible Conditioning Properties			Conditioned Property
	A	B	C	D	E
Occurrence*	Present	Absent	Present	Present	Present
1	Absent	Absent	Absent	Present	Absent
2	Present	Absent	Absent	Absent	Absent

If one of the PCPs is a sufficient condition for E in Occurrence*, then C is that sufficient condition. Note that B may be a sufficient condition for E, but B was not present on Occurrence*, and so could not have been a sufficient condition for E on that occurrence.

14.3.4 Combined Methods

The direct method of agreement (to find necessary conditions) may be combined with either the Inverse Method of Agreement or the Method of Difference (to find sufficient conditions) to find conditions both necessary and sufficient.

The *Double Method of Agreement* combines the Direct Method of Agreement with the Inverse Method of Agreement to determine a property both necessary and sufficient for E. First eliminate PCPs that are absent when E is present as necessary conditions. So if any of the PCPs are necessary conditions, they are the remaining ones. Then eliminate the PCPs that are present when E is absent as sufficient conditions. So if any of the PCPs are sufficient conditions, they are the ones remaining. But since those remaining have also passed as necessary conditions, they are both necessary and sufficient conditions for E.

Factor Tabulation				
Instance	Possible Conditioning Properties			Conditioned Property
	A	B	C	D	E
1	Absent	Present	Present	Present	Present
2	Present	Absent	Present	Absent	Present
3	Present	Absent	Absent	Absent	Absent
4	Absent	Present	Absent	Present	Absent
If one of the PCPs is a necessary condition for E and if one of the PCPs is a sufficient condition for E, then C is that property which is both necessary and sufficient.				

The *Joint Method of Agreement and Difference* combines the Direct Method of Agreement with the Method of Difference to determine a property both necessary and sufficient for E.

First apply the Method of Difference to a particular occurrence of E: (a) any property not present at Occurrence* cannot be a sufficient condition for that occurrence of E; (b) any property present when E is absent cannot be a sufficient condition for E for any occurrence of E. The properties remaining are sufficient conditions for E if any properties among the PCPs are.

Then apply the Direct Method of Agreement to eliminate as a necessary condition any property that is absent when E is present to obtain any necessary conditions for E.

Finally, combining the two results, any property that remains is both necessary and sufficient for E if any property among the PCPs is sufficient for E and if any property among the PCPs is necessary for E.

Factor Tabulation				
Instance	Possible Conditioning Properties			Conditioned Property
	A	B	C	D	E
Occurrence*	Present	Absent	Present	Absent	Present
1	Present	Absent	Absent	Absent	Absent
2	Absent	Present	Present	Present	Present

If one of the PCPs is present in Occurrence* is a sufficient condition for E, and if one of the PCPs is a necessary condition for E, then C is both necessary and sufficient for E. Note that B and D might be both necessary and sufficient for E, but they were eliminated because they were not present on Occurrence*.

14.3.5 The Method of Residues

The Method of Residues separates from a group of causal conditions (sufficient or necessary) those that are already known, leaving the remaining condition (the residue) as the causal condition of E. For example, suppose that the properties ABC are causally connected with the complex of properties $E_1E_2E_3$, and it is known that A is a causal condition of E_1 and that B is a causal condition of E_2. It is inferred that C is a causal condition of E_3 by the Method of Residues. Like the earlier methods, the Method of Residues is a method that involves elimination of all but one factor.

14.3.6 The Method of Concomitant Variation

The Method of Concomitant Variation serves to identify a causal condition (necessary or sufficient) between a conditioned property E and some PCPs by matching variations in one of the PCPs with variations in E. The variations may be direct or may be inverse. In such a case, the conclusion is drawn that if the PCPs include the causal condition, it is the one whose variation changes regularly with changes in E. In the table, "+" and "-" are used to indicate that a property is present to a greater or lesser degree.

Factor Tabulation			
Instance	Possible Conditioning Properties		Conditioned Property
	A	B	C	E
1	Present+	Present-	Present	Present
2	Present-	Present	Present+	Present+
3	Present	Present+	Present-	Present-

If a casual condition is present among the PCPs, it is C, which varies directly with E.

Factor Tabulation			
Instance	Possible Conditioning Properties		Conditioned Property
	A	B	C	E
1	Present+	Present-	Present	Present
2	Present-	Present	Present-	Present+
3	Present	Present+	Present+	Present-

If a casual condition is present among the PCPs, it is C, which varies inversely with E.

14.4 Probability

14.4.1 Theories of Probability

Truth tables provide a method of determining the truth value of a compound statement given the **TV**s of its atomic components, but gave us no method for determining the **TV**s of the atomic constituents themselves. In the same way, the probability calculus provides a way to determine the probability of a compound statement given the probabilities of its constituents, but offers no way to determine the probabilities of those simple constituents themselves.

There are at least three different theories of probability, each of which provides a way to determine the probability of simple statements. Each theory applies in somewhat different situations.

The *classical* theory of probability (or *logical* probability) determines the probability of a kind of event independently of any observations. The probability of an event E is determined by the formula

$\Pr(E) = f/n,$

where "f" is the number of favorable outcomes and "n" is the total number of possible outcomes. This theory assumes that all possible outcomes are equally probable.

The *relative frequency* theory assigns probabilities to simple statements on the basis of the frequencies with which certain kinds of events are observed to occur. The probability of an event E on the relative frequency theory is given by the formula

$\Pr(E) = f_0 / n_0,$

where "f_0" is the number of *observed* favorable outcomes and "n_0" is the total number of *observed* outcomes.

The *subjectivist* theory interprets probability in terms of the beliefs of individual people. The vagueness that attaches to this concept is given a quantitative interpretation via the odds a person is willing to take on a bet.

14.4.2 The Probability Calculus

The probability calculus codifies in a collection of rules and definitions what a quantity must satisfy in order to be considered a probability value. The probability calculus formulates the common core of each of the concepts of probability.

Probability values range from 0 to 1.

If a statement $\square$ is logically true, then $\Pr(\square) = 1$.

If a statement $\square$ is logically false, then $\Pr(\square) = 0$.

If a statement $\square$ is contingent, then $0 < \Pr(\square) < 1$.

If statements $\square$, Δ are logically equivalent, then $\Pr(\square) = \Pr(\Delta)$.

Statements $\square$, Δ are *mutually exclusive* iff $\{\square, \Delta\}$ is inconsistent.

<u>Negation Rule</u>: $\Pr(\sim\square) = 1 \; \Pr(\square)$.

141

Conditional probability is the probability that one event, Δ, will occur given that another event, $\square$, has occurred, $\Pr(\Delta \text{ given } \square)$:

Conditional probability: $\Pr(\Delta \text{ given } \square) = \Pr(\Delta \ \& \ \square) / \Pr(\square)$, where $\Pr(\square) > 0$.

Independence: $\square$ and Δ are independent iff $\Pr(\Delta \text{ given } \square) = \Pr(\Delta)$.

The *special conjunction rule* is used to calculate the probability that two events will occur when the events are independent of each other:

Special conjunction rule: If $\square$ and Δ are independent, then $\Pr(\square \ \& \ \Delta) = \Pr(\square) \times \Pr(\Delta)$.

The *general conjunction (disjunction) rule* is used to calculate the probability that either one or the other of two events will occur whether or not they are independent of each other.

General conjunction rule: $\Pr(\square \ \& \ \Delta) = \Pr(\square) \times \Pr(\Delta \text{ given } \square)$.

The *special disjunction rule* is used to calculate the probability that either of two mutually exclusive events will occur:

Special disjunction rule: If $\square$ and Δ are mutually exclusive, then $\Pr(\square \lor \Delta) = \Pr(\square) + \Pr(\Delta)$.

The *general disjunction rule* is used to calculate the probability that either one or the other of two events will occur whether or not they are mutually exclusive:

General disjunction rule: $\Pr(\square \lor \Delta) = \Pr(\square) + \Pr(\Delta) \ \Pr(\square \ \& \ \Delta)$.

Bayes' Theorem may be used to calculate the conditional probability of $\square$ given Δ from the probability of Δ given $\square$. For example, the probability that Lassie is a dog given that she is a collie is 1, but the probability that she's a collie given that she is a dog is less than 1.

The special version of Bayes' Theorem applies where events are mutually exclusive and jointly exhaustive, as Δ and $\sim\!\Delta$ are.

Bayes' Theorem:

$$Pr(\Delta \text{given } \square) = \frac{\{Pr(\Delta) \times Pr(\square \text{ given} \Delta)}{[Pr(\Delta) \times Pr(\square \text{ given } \Delta)] + [Pr(\Delta) \times Pr(\square \text{ given} \sim\!\Delta)]}$$

142

14.5 Statistical Reasoning

14.5.1 Samples

Many inductive inferences are based on analyzing samples. It is argued from the fact that if a sample has a certain property then the population as a whole has the property. For example:

93 percent of welfare recipients sampled were on welfare for less than four years. Therefore, 93 percent of all welfare recipients are on welfare for less than four years.

The strength of arguments based on samples depends on the sample being representative of the population sampled. A sample that is not representative is *biased*. Three considerations enter into whether a sample is biased: (1) whether the sample was randomly selected; (2) the size of the sample; and (3) psychological factors.

A sample is *random* iff every member of the population has an equal chance to be included in the sample.

Size plays an important role in determining how representative a sample is of the population. Given that a sample is randomly selected, the larger the sample, the more closely it conforms to the population. The degree of closeness is expressed as *sampling error*: the difference between the frequency with which a characteristic occurs in the sample and the frequency with which it occurs in the population. Roughly, the larger the population, the larger the sample size needs to be to attain a given degree of sampling error.

Psychological factors can have a bearing on whether a sample of a human population is representative. Such factors may arise, for example, when the people are questioned in a poll and think they may gain or lose something depending on the kind of answer they give; or the nature of the question may lead people to exaggerate or underestimate in their response to the questions; even the personal interaction between the person taking the survey and the respondent can affect the representativeness of the sample.

Sample Size and Sampling Error for Random Samples	
Sample Size	Margin of Error
4,000	±2%
1,500	±3%
1,000	±4%
750	±4%
600	±5%
400	±6%
200	±8%
100	±11%

14.5.2 The Meaning of "Average"

The word "average" is used in three different senses in statistics: as the median, the mean, and the mode.

The *median* value of a set of numbers is the midpoint of the group: exactly half the sample is above the median and half is below the median.

The *mean* value of a set of numbers is the number obtained by dividing the sum total of all the scores in the group by the number of individuals in the sample.

The *mode* value is the number which occurs the most frequently.

For example, suppose we have the following scores on an exam:

Student	Score
1	60
2	60
3	70
4	95
5	100

For this sample, the mode is 60, the median is 70, and the mean is $385 \div 5 = 77$.

14.5.3 Dispersion, Variance, and Standard Deviation

The normal probability curve is the familiar bell-shaped curve.

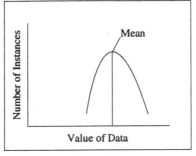

Figure 14-1

The *dispersion* of a set of data refers to how spread out the data are in regard to the value of the data; the bottom of the curve may be wider or narrower, depending on how much or little the values differ from the mean value. The *range* of values of a set of data is the difference between the smallest and the largest values. The *variance* and the *standard deviation* measure how far the data vary from the mean value.

For example, suppose a random experiment involving 10 repetitions yields the following values:

{4, 6, 8, 8, 8, 8, 9, 9, 10, 10}

The mean is 8. The sample values are scattered about the mean: the range of values is from 4 to 10. If we list the size of the deviation of each value from the mean we would obtain

{4, 2, 0, 0, 0, 0, 1, 1, 2, 2}

If we average these deviations, we obtain 1.2, the mean deviation from the mean. However, the first two deviations were negative, but the mean deviation from the mean uses only the magnitudes of the deviations and not their direction. Negative numbers can be avoided by squaring the deviations. The *variance* is the mean of the squares of the deviations from the mean. For the example above, the variance $v = (16 + 4 + 0 + 0 + 0 + 0 + 1 + 1 + 4 + 4)/ 10 = 3$. The deviations are measured in the same units as the values (if the values are grams, then the deviations are in grams). The variance is measured in those units squared. So to obtain a measure of deviation in terms of the units, rather than the units sqared, take the positive square root of the variance; this yields the *standard deviation*. For the example, the standard deviation = $\sqrt{3}$ = 1.732.